Ecology and Politics

ECOLOGY AND POLITICS

Environmental Stress and Security in Africa

Edited by

Anders Hjort af Ornäs
and
M.A. Mohamed Salih

Scandinavian Institute of African Studies, 1989

Cover: Details from a decorated gourd.
From *Nigeria's Traditional Crafts* by Alison Hodge.

Typesetting: Susanne Ljung

ISBN 91-7106-295-5

Printed in Sweden by
Motala Grafiska, Motala, 1989

Contents

Foreword

The essays in this volume are elaborated from papers presented at a working group meeting arranged by the Scandinavian Institute of African Studies for the conference on "Environmental Stress and Security", December 13–15, 1988. The conference was organized by The Royal Swedish Academy of Sciences, Stockholm, Sweden. Our interpretation of the theme was such that we focussed the working group activities on the theme "Environmental stress and political conflict in Africa". The contribution to the conference was that we reached beyond the issue of environmental degradation and political ways to deal with it, through development policy. The group sought to address the more precise issue of how environment and politics interrelate. Most contributions in the present volume follow this theme also.

We wish to express our gratitude to the Swedish Ministry for Energy and Environment for financing the early preparations for the workshop. Our sincerest thanks are due to The Royal Swedish Academy of Sciences for meeting travel costs for the participants of this particular working group.

Anders Hjort af Ornäs   M. A. Mohamed Salih

Introduction
Ecology and Politics

Environmental Stress and Security in Africa

Anders Hjort af Ornäs and M.A. Mohamed Salih

BACKGROUND

Recent African history has demonstrated how natural disasters such as drought may lead to starvation and disaster, sometimes fueled by an obvious conflict over land or access to food. A natural disaster may then have the effect to accentuate an inherent conflict over resource availability. One aspect of the correlation between ecological stress and political conflict is the issue of control over food. The quest for food security motors political conflicts to the extent that scarcity prevails, and causes ecological degradation in areas where pressure on land does not really permit increased food production without structural changes in the production systems.

This study is not a call for another penetration of conflicting management systems or the like. Rather, it concerns the twofold process where ecological stress leads to political conflict and vice versa. This way of relating environmental and political problems could enlighten us about issues which influence security both for nations and individuals.

The aim of the current collection of essays is in general terms to look into situations in Africa with a particular eye on how environmental and political issues interrelate. The assumption is, of course, that both these issues are of highest relevance for a comprehension of the total security situation. But what we have sought to achieve is a search for relationships between environment and politics. Thus, we have not been satisfied with just noting for instance a case of land degradation, and then search for the policy measures to counter such a situation. We have intended to go one step further: What kinds of environmental problems lead to political imbalance or conflict? And vice versa: Political conflicts which cause ecological degradation.

Africa is a continent, and naturally the contributions will not be able to produce a comprehensive map over the current situation vis-à-vis ecological stress and political conflict. Our approach when inviting participants has been firstly, to seek contributions on a few key issues with a continental or regional focus. Secondly, we have invited contributions from a number of country cases which we have felt could be significant for the topic at hand.

At the onset of the preparations of this volume we highlighted that person to whom food is not redistributed. He or she is the one who becomes obliged to wear on the land; hence man-made degradation. He or she is also the one causing political uncertainty, leading a dissatisfying life with a close margin to being forced to leave his rural context.

Peoples' or groups' absence of involvement in food redistribution systems is an indicator of a situation of potential conflict. Obviously an urban population, often the main concern in connection with political conflict, fits this picture; if food supplies from rural areas or abroad ceases, then unrest is at hand. But also that part of the rural population becomes directly concerned which is not involved in reliable food distribution; farm hands, peasants with part-time farming and occasional wage earning, small-scale traders, etc. This category makes up as much as 20-30% of the rural population in parts of Africa. The threat to them is starvation and/or migration to an urban slum. People in this situation become the carriers of both ecological and political problems, generated through food shortages. Their situation is in focus when we wish to comprehend linkages between ecological stress and political conflict.

Problems pertaining to the failure of appropriate development efforts in the arid lands are largely attributed to the lack of long-term policy formulations. Political effects of resource reallocation as a consequence of development efforts after ecological stress have been particularly overlooked. Consider for example the problems resulting from population concentration around newly constructed reservoirs, boreholes and settlement schemes. Another example is the mass migration of impoverished peasants and pastoralists to towns and centres of employment. Yet an example is the movement of some ethnic groups outside their territories to richer ecological zones already occupied by other groups.

Any inclination towards short-term policy regulations in such situations precipitate latent conflicts in the unforeseeable future. The same applies to internal refugees (those who inhabit the squatter settlements in the outskirts of large urban centres) or international refugees (those who crossed the borders to other countries for political or ecological reasons).

The general scope of a comprehensive approach to the problem of the association between ecological stress and political conflicts is wide ranging. The fact that political conflict easily leads to degradation is well established. The extent to which the reverse is true, however, is more difficult to establish. In order to formulate the qualities in the interlinkeages between ecological degradation and political conflict an interdisciplinary effort is required. We have intended to approach the problematique departing from a series of cases along with raising specific problems from local, national and regional levels.

The project has drawn on regional and country research experiences from Eastern, Western and Southern Africa with a number of specific country papers. These provide cases of serious ecological and political problems as well as situations where people or groups are on the verge of experiencing increased vulnerability from ecological stress. We might hypothesize that the security of the individual and that of the state at times do not coincide. The basic aim is to formulate viable solutions for man to maintain a sustainable livelihood without endangering national securities.

Conventional politically and ideologically motivated models are not sufficient to redress the problems of the vulnerable sectors of the population. The concepts of entitlement and sustainability enter the picture since the issues of deprivation, impoverishment and the need for viable solutions are closely interrelated.

The linkage between political instability and ecological stress is found also at the grand level of Africa's regional and subregional conflicts which have been precipitated by ecological degradation. We have outlined a "survey" of this situation, since it very much forms the context for the current study; the notion of security, for instance, is more often employed here than for the individuals concerned by the power field between ecological stress and political conflict. This contribution offers a geopolitical map of countries vulnerable to ecological stress which enhances political instability. As a corollary it also brings about the question of the "ecological refugees".

It follows that the main perspective of the study relates to a conscious focus on the mutual dependency, if any between ecological stress and political conflict. The perspective used is that of vulnerable individuals and groups and the security which these can establish for themselves parallel with the security of the state.

ISSUES ADDRESSED IN THIS VOLUME

In this brief introduction, we would merely like to provide a kind of appetizer of the topics discussed in the various contributions. All the papers presented in this volume relate to the topic how political conflict and environmental degradation may interrelate. In the following contributions a number of illustrations are given how conflicts lead to ecological stress; this can be a fairly straight forward case like during an armed conflict, or a more subtle process generated by a kind of centre/periphery relation. Also relations of the reversed direction are given; i.e. when ecological stress leads to political conflict. The most glaring example is of course that of ecological refugees to towns and cities in many African countries, people more or less pushed out of a rural context moving to a restless life on urban fringes.

The perspectives in the contributions vary. Some have a regional focus, while others have either national or local foci. This variation is also what we had hoped for according to the above section. It is extremely helpful when we wish to outline the complexity of the relational set-up, but also when we try to raise the question about scale invariance: Are there factors which operate on local, national as well as regional levels? An obvious example these days is how the chain of conditionality in World Bank financed projects has an effect at different levels; a regional policy of that institution leads to the formation of national policies, influencing both the countries' development aid cooperation and their rural development strategies, all leading to consequences for local farmers, for instance to change production focus from food (like sorghum) to more marketable products (like oil seed).

Answers to the question of how environment and politics interrelate on various levels will be extremely valuable for more strategic considerations, be it for development or other purposes. The conference where the contributions to the present volume were first presented highlighted "security", and it is our hope that the present volume can contribute significantly towards giving at least evidence for a wide notion of security as an appropriate approach in the African case. It is crucial that we try to stay away from rhetorics, whether using this label, "sustainable development" or some other to signify the complex needs of the world today. And to our minds one has to work just like we did in our sessions; try to probe some of the key issues in a fairly detailed manner on regional, national *and* local levels.

The reader will find that it would be possible to list fairly distinctly a number of relations between the poles environment and politics. Every author contributes towards forming a kind of web of both empirical cases and analytical perspectives. True, it is unusual to be able to point at one particular issue, and say that this is the cause of either environmental degradation or political conflict. But it seems feasible to point at a few crucial issues, which are highly significant.

To begin with the actual hypothesis inherent in the approach: There seems to be some sort of *correlation between environmental degradation and political conflict*. This assumption made at the onset of setting up this workshop is in essence verified in regional terms (Zdenek Červenka), and national evidence is given (Bekure Semait) for the case of Ethiopia. But in spite of a strong correlation the degree of causality is difficult to establish. The literature is clearly weak on this point, more research is called for, and the working group has a very significant issue to penetrate.

Resource management and changing conditions, be it for control over farm land or pastures, or over the produce, comprise themes which reoccur in many of the contributions. Both local cases (Carl Christiansson and Eva Tobisson, Anders Hjort af Ornäs) and national ones (Bekure W. Semait, Abdel Ghaffar Mohamed Ahmed, M. A. Mohamed Salih) illustrate the situation in Eastern Africa with rapidly changing resource management. In Southern Africa, too, it is noted (Kwezi K. Prah, W. P. Ezaza and Haroub Othman), albeit with the "complicating" fact of South Africa's destabilization policy.

The security issue is brought up on several levels. Again the Southern Africa situation has to be mentioned (W. P. Ezaza and Haroub Othman). Along with the Uganda case (Byarugaba Emansueto Foster) it demonstrates how *political conflicts cause environmental degradation*. But we do not only have national and regional cases. On the local scene we see illustrations from Lesotho (Kwesi K Prah) and Ogaden (John Markakis) with great time depth, from Kenya and Sudan with empirical evidence (Abdel Ghaffar M. Ahmed, Anders Hjort af Ornäs, M. A. Mohamed Salih), and from West Africa with the conclusion that local conflicts have a national cause in political boundaries with a regional solution (Okwudiba Nnoli).

The ecological stress is so aggravated today that a notion of *ecological refugees* is already well established. This issue is raised in many of the contributions to the workshop, notably for Uganda, Sudan and Ethiopia in Eastern Africa (Byarugaba Emansueto Foster, M. A. Mohamed Salih, Bekure Semait, Michael Ståhl), and countries like Mali, Niger, Nigeria, Chad and Cameroon (Okwudiba Nnoli) in West Africa. The situation is in several instances very serious and calls for much greater attention to *human rights* and other legal issues, such as judicial land rights.

The *demographic issues* raised in the contributions are several. Two contributors concentrate on this topic (Christer Krokfors, Norman Myers), but several papers touch upon population pressure as one very significant factor behind acute conflicts, adding to the complexity for instance in Southern Africa (Adolfo Mascarenhas) and Ethiopia (Michael Ståhl). The obvious case to relate overpopulation with land degradation goes via urbanization (Norman Myers, M. A. Mohamed Salih, Abdel Ghaffar M. Ahmed among others) and unemployment linked with pervasive poverty and refugee status. The demographic picture of Africa is still one of rapid growth, although modifications must be made; an increasing political operational awareness, and, in cases, a strikingly successful mixed policies of incentives and taxation to improve the situation (Norman Myers).

The role of the state is touched upon in most of the papers, either with a negative connotation, the state being repressive or absent in times of famine, such as in Sudan (M. A. Mohamed Salih) or with a more complex planning policy discussion.

Development strategy and planning is of course an issue raised in most papers. We then come back to the earlier topic of scale invariance; how local, national and regional problems have their roots on several levels. A few authors refer to "Our common future" more on the level of perspective; the problem of environmental degradation today, be it of land as in many African cases or else, can not be formulated in technical terms. This will lead us to symptoms of a problem, not to its roots. To select a sophisticated notion of security, beyond a military strategic one or one based on nations as the smallest entity, would seem to be a fruitful approach, judging from the contributions to the current workshop. We would then have to place "security" in a context; that of the individual and her social group, the state, the region, etc. Ecological stress may lead to political conflict of varying kind. The ways in which security is established, or not established, by and for people living off the land becomes a key topic.

THE CONTRIBUTIONS

The papers presented at the workshop on Ecological Stress and Political Conflicts in Africa' could be classified into regional, sub-regional, country experiences and local level studies. All have depicted the themes of the proposal but not necessarily agreed with its premises. This has enriched the scope of the debate and the issues raised.

Regional studies

Červenka's paper on the 'Relationship between Armed Conflicts and Environmental Degradation in Africa' offers an ample explanation of the current geopolitical situation in Africa. The paper draws on current material to establish a link between armed conflict and environmental degradation in the hot points of political confrontation in Africa. The theme has highlighted an all familiar crisis situation in which the poor and the needy have suffered most. The horrors of war and their impact on human misery, destitution and flight is beyond comprehension. This has been the situation in Angola, Mozambique, Burundi, Somalia, Ethiopia, the Sudan and the Western Sahara.

According to the author some of the main consequences of ecological stress, in addition to famine and starvation, is the suspension of development projects, loss of herds which represent the main wealth reserve for pastoralists, the erosion of morality and social solidarity, the deterioration of the status of women,.and the violation of human rights. Červenka proposes in accordance with the invitation to the workshop, that a new meaning of security has to be put forward. It should first and foremost include food, physical survival, family and community security rather than military security. An approach to this agenda should consider an embargo on the export of arms to Africa and the diversion of all resources to the realization of this alternative mode of security.

Ahmed's 'Ecological Degradation in the Sahel: The Political Dimension' summarizes the conventional reasons behind ecological stress and food shortage in the African Sahel. He argues that crisis is a consequence of the lack of capital resources, and know-how to deal with deforestation, desertification, drought and population growth. Ecological stress has furthermore been aggravated by other factors such as over-cultivation, overstocking and the recurrent civil wars. The failure of the present plans to remedy the situation is attributed to the implementation of inappropriate plans which failed to take into consideration indigenous environmental knowledge.

Furthermore, national Governments in the Sahel are operating under external pressures because they are assigned to the role of raw material producers. Contrary to the view which expresses the poverty of Africa, there has been a regular transfer of wealth from the poor Sahelian countries to the rich countries. In other words, the resources which could have been used for development or combating the ecological crisis have been transferred to finance the riches of the industrially advance rich countries. The author emphasized that most of the development plans seem to have no regards for the nomadic population and popular participation.

However, the author argues that blaming the whole issue of underdevelopment and ecological degradation on experts, foreign and national, is not sufficient. The local population and some of their practices which are detrimental to the environment have to be assessed too.

Myers' paper on 'Population Growth, Environmental Decline and Security Issues in sub-Saharan Africa' points out that the present three percent rate of population growth indicates that the population of Africa is destined to increase from 508 in 1988 to 678 in the year 2000. In a continent which imports one fifth of its cereal production, the food deficit is also destined to increase. A high demand on international river basins such as the Nile is anticipated to increase as countries such as Egypt, the Sudan and Ethiopia inclinations toward irrigated agriculture increase. This is also anticipated to increase the already mounting conflicts between these Nile basin countries. The Ethiopian plan to divert the Blue Nile as part of its ambitious irrigation programme would contribute to water shortages in the Sudan and Egypt. A decrease in the amount of water available for Sudan's huge irrigation schemes and Egypt dependence of Aswan Dam as the only source of irrigation signal a dangerous course in the relationship between these three countries. The same theme could be applied to Ethiopia/Somali dispute.

The recurrent famines from which Ethiopia and Somalia suffered in recent years, could be explained against environmental stress, high population growth and a high investment on the war effort. This has been followed by an unprecedented record of environmental refugees prompted by military activities, faulty development policies and ultra-rapid population growth.

The issue of population and ecology has also been tackled by Christer Krokfors, who in his paper 'Population and Land Degradation' argues that the politicalization of landed resources reflects the political structure of a given population. Vulnerability, land degradation and over-population all have to do with the political nature of resource allocation. People's inability to reap the development promised by the state has, in some cases, led to the creation of new ideologically based land use practices. Multi-active households, represent one form of secluded-group strategy with many socio-economic and political implications for development which are yet to be assessed within the confines of the present economic and ecological crisis.

Sub-regional papers

Nnoli's paper on 'Desertification, Refugees and Regional Conflicts in West Africa' elucidates the magnitude of desertification, drought and its disastrous impacts such as famine, shortage of water, drying rivers and reduced vegetation growth. The politico-social results of this process is mass flight of population to neighbouring countries and rural/rural and rural/urban migration. Desertification therefore has the potential of creating high population mobility which is also a source of conflict within and across nations. Such conflicts arrest development efforts and divert resources from economic and social planning priorities.

The fluid boundaries of the West African states facilitate the movement of refugees, but causes economic imbalances through smuggling of goods and food, and create competition over the meagre social and public services. "The Free Movement of Persons, Residence and

Establishment Treaty" represents a good intention in part of the West African states. What is needed now is a specific emphasis on refugee rights and the total removal of residence restrictions across international boundaries.

Nnoli argues that political conflicts in West Africa are not directly caused by refugees, but rather by ecological degradation represented by the desire of some states such as Niger, Nigeria, Chad and Cameroon to conserve more waters from the international river basins. For example a political row erupted when Niger constructed dams in Lamido and Kalmalo tributaries of the River Niger which affected water level in Nigeria. The resolution was that Niger should halt the construction of these dams while Nigeria can provide it with electricity. Another example is Chad's displeasure with Nigeria and Cameroon irrigation plans which affected the water level in Lake Chad. This has prompted the establishment of River Niger and Lake Chad Commissions to solve the disputes which might arise. The point to emphasize here is that desertification makes the West African states more dependent on international rivers which is a potential source of political conflict.

The situation in West Africa with respect to international water basins resembles that of East Africa which is described by Myers' analysis of the situation in the Sudan, Ethiopia and Egypt. At a regional perspective it is therefore very important to evaluate the interchangeability between the arid and semi-arid lands and the river basins and humid and sub-humid zone. The fear is that ecological stress, and subsequently disaster, lead to inter-state conflicts.

Mascarenhas has examined 'Environmental Stress and Security in Southern Africa' and offers a detailed description of the extent of ecological degradation in the SADCC countries. The main thesis is that the political instability caused by racist South Africa has aggravated ecological stress through destablization. Hence it created a dilemma for these countries which have the difficult problem of conservation, on one hand, and uncareful utilization of resources to maintain the present status-quo, on the other.

In Tanzania for example, Ngorongoro conservation park is a clear example of a conflict of interest between the planners and the Maasai over the objectives of the project. Efforts such as those commenced by Zambia and Zimbabwe by enacting National Conservation Strategies, reveal a great concern with ecological problems. Yet both countries have to overcome several structural and infrastructural difficulties before their efforts can be truly effective. Moreover, although Botswana is one of the richest Southern African countries, its environment-based economy requires more attention. Its arid and semi-arid environment is susceptible to drought and irreversible environmental degradation.

The case of Mozambique demonstrates how destablization leads to population concentration, distortion of traditional sustainability, lack of rehabilitation and excessive utilization of the environment. The main premises of Mascarenhas' paper is that conservation is the gate to security because the only assets which are commonly distributed throughout the rural and generally poorer sectors of the population, are the natural resources. A sustained improvement in the standards of living is impossible without conservation. The biggest threat to subsistence, the environment and development in the SADCC countries is the detribalization policies of apartheid South Africa which disastrously cut through their economic and political structures as well as the natural resources and the environment.

Ezaza and Othman consider the relationship between 'Political Instability and Ecological Stress in Eastern Africa'. The paper deals mainly with food insecurity in Uganda and Mozambique and its linkages to political instability represented by war and ecological instability. Ecological degradation are explained as a result of the degradation of land, vegetation, climate and water resources. During the wars, agricultural production was disrupted, and resources are indiscriminately used. According to the authors, if the present rate of deforestation continues, there will be no pure forest stands in Uganda by the end of this century. This is attributed mainly to the fact that swamps are drying up by being over-used by brick-makers who devastated the forests and the vegetation cover.

The case of Mozambique is described as starvation by design in which not only racist South Africa is blamed but also Western investment which helped South Africa to destabilize the country. Political and economic destabilization causes ecological imbalances through the disruption of production activities and population concentration which contributes to an uneven utilization of resources.

Ezaza and Othman conclude that the war in Uganda has precipitated national and international conflicts across the borders with thousands of Ugandan refugees living in camps in the Sudan, Kenya and Zaire. Many Mozambicans have fled their country to Malawi, Botswana, Zambia, and Zimbabwe. If wars and political instability continue, they will be followed by ecological degradation which would in turn cause social unrest. Ecological stress and political conflicts are therefore two sides of the same coin.

Country experiences

Two papers dealing with Ethiopia are presented by Semait and Ståhl. Semait's paper is on 'Ecological Stress and political Conflict in Africa:

the Case of Ethiopia'. It provides a brief, but systematic explanation of the nature and characteristics of the political conflicts in Eritrea, the Ogaden and Tigray. Thence the author proceeds to mapping ecological stress in terms of the deterioration in the climate, rainfall, vegetation cover and high population growth of over 3%. According to Semait, spatial coincidence between ecological stress and political conflicts has been established, but a precise cause-effect relationship is difficult to subscribe. The author argues that the settlement programme has been discredited by the Western mass media and Ethiopian refugees living in the West as part of their campaign against the Ethiopian revolution. However, such tendencies may transform the settlements into a squabbling ground between old timers and new comers. The potential of acute political conflicts resulting from the settlement programme can be fueled by the Western powers' desire to perpetuate a situation for undermining the whole project.

The second paper on the Ethiopian case is presented by Ståhl under the title, 'Environmental Degradation and Political Constraints in Ethiopia'. After an introduction on development and ecological degradation the author moves into a discussion of the factors which contributed to the emergence of the nationality question and the liberation movements. The contribution establishes a correlation between the evolution of the state society, political degradation and ecological stress. The author comments that erosion accelerates in the wake of the political stalemate between the Government, the opposition groups, the peasants and the donors. Vegetation is relentlessly deteriorated by grazing livestock and by humans in search of firewood and building material. The ability of the peasants to withstand periodic droughts decreases. This view runs counter to Semait's view that the Western Governments and donors impose their prescriptions as how the Ethiopians should govern their country and that they in doing so discredit any Ethiopian move to address the problems of development and rehabilitation. Ståhl's paper demonstrates clearly the type of conditionality that international donors impose on aid receiving countries whether for development *per se* or for the rehabilitation of a degraded ecology.

In 'Land Degradation and Class Struggle in Rural Lesotho' Prah argues that despite labour migration to South Africa, rural Lesotho is over-populated in terms of the criteria of allocation of land with 20.7% (i.e. 45,549) of the total population landless. The stocking rate of grazing animals is 300% higher than the land's carrying capacity. Over-grazing and poor conservation measures have resulted in low crop productivity. Amid these facts the inherited colonial land tenure system has been preserved until the Land Act of 1979 which represents a compromise between chiefly and feudal interests. As such it removed from rural production its key factors of production; labour. It has at the same time exposed land to marginalization and the retrogression of its productive forces. The class conflict here is between the small producers, on the one hand, and the traditional leaders, bureaucrats, and landlords who use the state apparatus to ensure their access to capital resources. The main argument is that the dominant classes are concerned with maximizing profit from land with little or no emphasis on conservation. The end result is land degradation and a conflict of interest between the landless and the landlords.

The paper by Salih on 'Ecological Stress, Political Coercion and the Limits of State Intervention; Sudan' is closely associated with the paper by Ståhl. The paper outlines the intervention of the Sudanese state in the ecology and the consequences of this process. Ecological degradation is attributed to the establishment of large-scale mechanized schemes, population and livestock concentration around water sources and the evolution of large urban centres with high demand for firewood and charcoal. This is coupled with the use of coercive measures to displace traditional farmers and pastoralists from their lands in favour of large-scale agricultural mechanization. This process has created several conflicts between the owners of the large-scale mechanized schemes, the cultivators and the pastoralists.

The state has also used coercive measures to restrict the movement of pastoralists and farmers, during the 1983/1985 drought phase and prevented them from settling in the wetter zone. The state has also instituted the *kasha* (i.e. unwarranted detention) which facilitated the repatriation of the victims of famine to their ecologically degraded areas instead of encouraging spontaneous settlement. Due to underdevelopment, the state objectives have witnessed double retreat from development to crisis management and from crisis management to bare maintenance of order and compliance. The state inability to intervene on behalf of the victims of drought and famine has lead to the use of coercion as the only means to legitimize its holding of power and to justify its faulty distributive mechanism.

Foster's paper elucidates the case of Rwandese Refugees in Uganda and their impact on the environment. Political conflicts between the rivalry Batutsi and Bahutu usually contribute to the flight of the Batutsi to Uganda. The influx of the Rwandese exert pressure on the environment and negatively affects the social setting of the population in Uganda/Rwanda border. The rural population in Uganda have to be squeezed or forced to move to other parts of the country in the face of the influx of Rwandese refugees. Moreover, the refugees have in turn created conflicts between the refugees and the local populations through their involvement in national politics in Uganda during the civil wars. The Alien Registration and Control Bill is specifically designed by the Government of Uganda to curtail the influx of refugees rather than to help integration. It has in fact encouraged the local population to harass Rwandese refugees and seize their property. The case of the Rwandese refugees in Uganda illustrates the intimate relationship between political conflicts and ecological degradation.

In contrast to the Treaty of the Free Movement of Persons, Residence and Establishments, the abolition of visa requirements for West Africans traveling in West Africa, many East African states, notably Sudan, Uganda, Somalia, Tanzania and Kenya have introduced further restrictions on the movement of refugees. This is despite the fact that over 50 percent of the African refugees come from East Africa. The proposition put forward by Nnoli which advocates the removal of movement restrictions between states seems more appropriate for the East African situation.

Local level studies

Markakis' paper on the 'Ishaq-Ogaden Dispute', relates the history of land dispute between the Ishaq, the Ogaden, the Somali state, Ethiopia and the Italian colonial regime. The worsening ecological situation in the 1940s led many pastoralists to migrate into Ogaden. The concentration of large numbers of livestock around the water sources contributed to overstocking and overgrazing. Congestion led to increased violence and disputes over grazing and watering rights among the various clans. The situation was aggravated by the opening of large livestock market in the oil rich countries during the 1960s along with the Ethiopian/Somali Ogaden war in 1977 and the concomitant proliferation of various opposition groups. At the national scene, the Ishaq were armed by the Somali National Movement while the Ogaden were supported by the Western Somali Liberation Front. As a result the pastoralists have been squeezed between two waring parties and hundreds of thousands were forced to leave the homelands only to become destitute living on International charity.

In their contribution on 'Environmental Degradation as a Consequence of Socio-Political Conflict in Eastern Mara Region, Tanzania', Christiansson and Tobisson argue that scarcity of water, dwindling timber and firewood resources, increase in grazing cattle raiding, and risky and unreliable agricultural potential have acted as catalysts and ignited socio-political conflicts among local population in the Eastern Mara region of Tanzania. In other words ecological stress is directly related to local conflicts. The conflict situation also militated the need for strong fencing and created an extra demand on the forests. The end result is a widespread ecological degradation.

21

Hjort af Ornäs' paper on 'Environment and Security of Dryland Herders in Eastern Africa' examines local case studies from northern Sudan and northern Kenya. The balance between family herd and size of household are considered the most crucial aspects of pastoral viability. In this situation security is achieved through dependence on relatives, borrowing of animals, redistribution through marriage etc. Risk spreading is seen as one of the general principles of social behaviour. The two cases, however, provide different rooms for manoeuvre depending on their resource base. The author concludes that the political connection with ecological stress is primarily on the local rather than the national political scene.

The paper is a call for a new notion of security, including the community based systems for resource management. Only by establishing secure access to food on the local level can a sustainable development be established. And without it, not only local but also national and international conflicts draw close to the extent that international security is threatened.

CONCLUDING REMARKS

This study concerns the interrelationship between ecological stress and political conflicts. We may speak about a geography of African political conflicts and relate it to ecological problems in regional terms. On sub-regional levels we can deal with boundary problems and access to natural resources which transcend several countries, not least river basins. In national terms we may relate specific political conflicts with land use more rigidly, introducing rural inhabitants as a social class. On the Local level we may come closer to resource management and conflicts over access to sparse assets. These four levels of analysis do not operate independently from each other and that they are essential in circumventing any tendency towards exalt one level of analysis at the expense of the others.

It becomes important to conclude all levels in a proper analysis of environment and security. We cannot be satisfied with one or few interpretations what amounts to a complex situation. And it does not help with an ecological interpretation, for instance: decision-making and the interests of populations influence how natural resources are utilized. In other words, there is a political dimension to environmental management. Conflicts and ecological stress are, therefore, interrelated. The issue deserves further attention, since future key issues for Africa certainly include the problem of 'ecological refugees' and sustainable usages of natural resources, adding to the more conventional security issues which relate to political boundaries and assets shared by several countries. This volume has, therefore, depicted the general pattern of the interrelationship between ecology and politics. More indepth studies relating the regional, sub-regional, national and the local levels are badly needed to colour in the details of this process and its future trends.

The Relationship Between Armed Conflict and Environmental Degradation in Africa

Zdenek Červenka

AFRICA'S WARS IN 1988

Most of the ongoing armed conflicts in Africa have a long history and origins which in many cases go back into pre-colonial times. Even a brief survey of the causes of Africa's wars would exceed the scope of this paper which is addressed to their consequences only. In order to illustrate the magnitude of the devastation brought about by men in arms on their fellow citizens and their environment the following is a brief country survey of armed conflicts in Africa in 1988. This is not the place for examining the causes of the conflicts but to show their consequenses.

INTERNATIONAL CONFLICT IN ANGOLA

In 1988 the war in Angola went into its thirteenth year. It is impossible to establish the economic losses attributable to the war conditions suffered without respite since independence in 1975, but they are clearly enormous. They have been incurred at several levels: direct damage resulting from the hostilities, the disruption to economic activities, the diversion of government revenue and foreign exchange resources to military expenditures estimated to be in the region of well over 40 percent of total national income. Scarce skills to have been diverted to the armed forces.[1] The exact number of the people who died in the war or of its consequences will probably never be known but it runs into hundreds of thousands of lost lives. One indication of the proportion of the human tragedy are 20,000 persons, many of them children who were maimed by mines planted in the roads by the combatant parties. More than 700,000 people fled the most fertile lands of the central highlands, a traditional "bread basket" of Angola which was virtually destroyed by the fast growing jungle. These "deslocados" who once were producers of food were saved from starvation only by a massive international relief effort.

Although South Africa withdrew its armed forces from Angola and an Agreement on Angola providing for the withdrawal of the Cuban troops was signed on December 22, 1988 in New York, the war between the guerrillas of UNITA and the Angolan army still goes on, thus leaving the recovery of Angola's economy and restoration of normal conditions of people in rural areas to an uncertain future.[1]

THE DESTRUCTION OF RURAL LIFE IN MOZAMBIQUE

The destruction inflicted upon Mozambique's rural areas by the rebel forces of RENAMO reached its peak in 1988. The situation was described by Ray Stacy, the US Deputy Assistant Secretary of State for African Affairs, as "one of the most brutal holocausts against human-beings since the World War II"[2]. It is estimated that at least 100,000 Mozambicans were killed by RENAMO forces backed by South Africa. Perhaps the most graphic indicator of the traumatic conditions in Mozambique is the fact that 3.3 mn people, that is 23 percent of the country's population of 14.6 mn, face severe shortage of food and other items essential to survival. Of these, 1.1 mn have been uprooted from their farms and homes and have lost their means of subsistence and almost all their possessions. It is estimated that an additional 2.6 mn people in rural and urban areas are affected by commercial food shortages. And, in addition to the 1.1 mn internally displaced persons, 700,000 Mozambicans have fled to neighbouring countries to escape the war, more than 400,000 of them to Malawi alone.

THE RECURRENCE GENOCIDE IN BURUNDI

In August 1988 there was a recurrence of the 1972 genocide killings in which 100,000 Hutus lost their lives. This time, the Hutus, who constitute 85 percent of the population, revolted against the ruling Tutsis in protest against Tutsi army operations on the border with Rwanda which attempted to stop the smuggling of coffee, a crop providing more than 80 percent of the Hutus' income. About 2,000 Tutsi soldiers were killed by angry Hutu crowds and their bodies were dumped into the rivers. The Tutsi army retaliated by using helicopters, grenades, machine-guns and rifles against machetes and killing about 10 times as many Hutus. About 50,000 refugees fled accross the papyrus swamps to Rwanda thus adding to the serious economic problems of this most densely populated African country. They left behind them an eerie, near empty land of destroyed villages and fertile land returning to a jungle. The recovery of the region emptied of its population is beyond the means of the Burundi Government.[3]

A SHORT-LIVED OUTBREAK OF CIVIL WAR IN SOMALIA

On May 27, 1988 the rebel forces of the Somali National Movement (SNM) surged across the frontier from their camps in Ethiopia and occupied parts of the city of Burao, about 130 km southeast of Berbera, and of Hargeisa, former capital of British Somaliland. The rebels were gambling that their assaults would bring about the collapse of the Government of President Siad Barre of Somalia. After three months of fighting they were defeated and both Burao and Hargeisa were reduced to rubble by Somali artillery and bombing. Reports speak of 35,000 people, mostly women and children, fleeing into the Ogaden region of Ethiopia, and another 40,000 into the Somali countryside. Amnesty International accused the Somali Government of the systematic massacre of civilians. It described the killings as revenge committed by soldiers who tortured and executed arrested people and as part of a persistent pattern of gross human rights violations by the Somali security forces. The areas where the fighting took place were sealed to visitors so that the amount of damages suffered by people and the destruction of their environment is difficult to assess and impossible to rectify.[4]

NEW KILLINGS AND DESTRUCTION IN THE FORGOTTEN WARS OF ETHIOPIA

Surprisingly little world attention is being paid to the deepening and dangerous crisis unfolding in Ethiopia where there is the twenty-eight years long war between the Eritrean People.s Liberation Army (EPLA) and the Ethiopian Government Forces. It took a new turn in March 1988 when the EPLA won its biggest military victory, though at high costs of lives and human misery. The victims of this intractable conflict are millions of Ethiopians who have already died from famine or been scarred by their experience, and the hundreds of thousands of other civilians who have been killed, wounded or uprooted.

EPLA advances continued throughout the year, killing hundreds of Government troops, destroying tanks and vehicles and seizing large quantities of arms. Its offensive was stopped only by massive bombings by the Ethiopian air force and deployment of fresh troops recruited under an emergency mobilization plan. The fighting had considerably aggravated the ecological crisis in Ethiopia which experienced the most serious droughts and famine in the continent.

Equally successful was the offensive of the Tigray People's Liberation Front, fighting for control of Ethiopia's northern province. During its offensive in March 1988 the Front captured the towns situated on the main highways and gained control of the strategically important road connecting Addis Ababa to Asmara in neighbouring Eritrea. After the capture of the garrison town of Maychew, south of the provincial capital of Makele, it forced the Ethiopian army to withdraw from southern Tigray into Wollo province.

The response of the Addis Ababa regime to its military setbacks in Eritrea and Tigray was a new mobilization of conscript irregulars and a resolve to commit even more of Ethiopia.s scarce resources to the war front. According to a report published in the Observer of June 5, 1988 the new conscripts include 14-year-old children who were snatched from a playground and hauled off to army camps for a short spell of military training before being sent to the killing fields of the Eritrean front. They have taken their place among an estimated 16,000 Ethiopian soldiers who were forcefully recruited and kept in the army, some of them for more than a decade. Ethiopia decided to cut off the outside world from the news of Ethiopia's military defeats and of the people in the area of fighting affected by severe lack of food supplies. President Mengistu ordered all voluntary aid agencies, except for the Save the Children Fund, to withdraw their relief teams from the famine-stricken areas of Eritrea and Tigray.[5]

THE WAR WOUNDS IN SUDAN

The civil war in Sudan has now been going on with varying intensity for the greater part of the three decades since the country achieved independence in January 1956. On November 8, 1988 it was described by Christopher Patten, the British Minister for Overseas Development Aid, as one of the cruellest, yet almost forgotten wars of our time.[6] In 1988, the Sudan People's Liberation Army (SPLA) continued to be engaged in heavy fighting with the Sudanese army in the new upsurge of the civil war renewed in 1983.

The conflict wore the rural economy of southern Sudan to shreds, and millions of families face the choice of either attempting a perilous 1,500 km journey to the urban squalor of Khartoum or a three-month trek to refugee camps in drought-stricken Ethiopia. The situation in southern Sudan is well illustrated by the following quote from "War Wounds", published in October 1988—one of the most moving accounts of the war:

"The civilian population in south Sudan see four enemies, all of them deadly: government troops, the Sudan People's Liberation Army (SPLA), the tribal militias,the famine. And they are defenceless against any of these.This is the state of affairs in every corner of the south today. Killing and looting are prominent. Civilian casualties have far outnumbered those among the armed forces. A veritable genocide is being perpetrated...About one million people are in Khartoum, and western Sudan and in Darfur and Kordofan provinces, after fleeing from the south. Hundreds of thousands of displaced people cling to camps on the edges of towns in the south. Similar numbers are found across the borders in Ethiopia and Uganda".[7]

In August 1988 the ravages of the war were overshadowed by a natural catastrophe of biblical proportions: Sudan experienced the worst floods of this century with one and half million people losing their homes and half of the capital of Khartoum washed away. The war in Sudan has cost more than one-and-a-half million lives and produced over three million refugees.

OTHER AREAS OF ARMED CONFLICTS

Civil war has not yet ceased in Uganda. The Museveni regime has the unique distinction of using the term "massacre" to describe its military successes over its armed opponents, the Uganda People Democratic Party (UDPA) and the Uganda People.s Army (UPA). The Ugandan National Revolutionary Army.s (NRA) communiques regularly announce that anything from 100 to 400 rebels have been "massacred" in fighting. There is no evidence of prisoners being taken. These "massacres" are not only confined to military engagements, there are well-substantiated reports of civilians being rounded up in villages. They are then either killed or arrested as suspected sympathisers of the rebels. The difficulty of establishing the true situation in the Northern and Eastern parts of Uganda is that the regime has forbidden diplomats, journalists and human rights organizations from traveling to parts of Acholi, Karamajong, Lango and some of the Eastern districts.[8]

A tenuous ceasefire reigns in Chad where the 20–year-long war ended in September 1987. Fighting was also temporarily suspended in the ten-year-old struggle of the POLISARIO for the independence of Western Sahara.

Perhaps one of the most dangerous potential conflicts is simmering in South Africa where tension between the blacks and apartheid regime is mounting. The ANC resorted to armed attacks against civilian targets and South African police incited violent ethnic conflicts. South African armed forces regularly commit acts of aggression against Zambia, Botswana and Lesotho in hot pursuit of members of ANC and its armed wing. Failure of the plan for Namibian independence agreed upon by the US, South Africa, Angola and Cuba is bound to lead to the escalation of fighting between SWAPO guerrillas and the South African occupation forces in Namibia.

All these wars have been leaving a blazing trail of abandoned villages littered with corpses, scorched homesteads, burned out crops, large patches of land stripped of tree cover and depleted food and water resources. In the Horn of Africa, in particular in Ethiopia and Somalia as well as in the Sudan, the wars contributed to an already serious degradation of the environment, posing a serious threat to the lives and safety of families and whole villages. The combined effects of wars, drought and desertification cause millions of people to set off in search for food and shelter.

THE INTER-RELATIONSHIP OF ARMED CONFLICTS AND ENVIRONMENTAL DEGRADATION

The direct consequences of armed conflicts on the regions of Africa under ecological stress, which are most pronounced in the Horn of Africa, can be summarized as follows:

Suspension of development projects

Armed conflict leads to a virtual suspension of development projects, some of them vitally important for redressing the causes of drought and desertification. For example, all development projects in southern Sudan are at a standstill, including the Jonglei canal scheme initiated in 1978 and the development of oil deposits known to have an initial capacity of 50,000 barrels per day.

Destruction of cattle herds—insurance for the survival of pastoral people

Livestock are crucial in the socio-economic structure of the pastoralists. For survival in their marginal environment, the pastoralists keep large herds of cattle as a form of insurance against natural environmental and man-made hazards, such as lack of foods, epidemic diseases, ethnic conflicts and civil wars. Herd sizes vary from ten to several hundred head. When there is a drought resulting in grain shortage, cattle, as well as sheep and goats, are sold or exchanged for grain. Social and cultural interactions relating to marriages, ritual settlement of disputes also entail possession and disposal of large numbers of cattle. The tendency to retain large herds of cattle is a universal phenomenon in pastoralist Africa. The destruction of the herds constitutes not only an economic loss but, more significantly, disrupts the fabric of the life of pastoral people.

Erosion of morality

The following example shows how war causes the loss of the sense of value and dignity of human life, and a growing trend towards violence:

In July 1987, 30 displaced people, including women and children, were gunned down at Nesitu, southern Sudan, about 20 km from Juba, apparently by SPLA rebels. Several others were seriously injured and had to be admitted to Juba Teaching Hospital. They told, how rebels without provocation, bombarded the camp for half an hour with rocket-propelled guns and handgrenades. They said that the 11 policemen guarding the camp ran away as soon as the shooting started. When the rebels found there was no response, they rampaged into the camp killing, looting and destroying tents. They said the SPLA told them they were to be killed to show the government that the situation was "alarming".

But the SPLA should not be singled out for condemnation alone. In Wau, on 11 and 12 August 1987, the army attacked three residential quarters and killed at least 600 people, mainly women and children. In April 1988, in Juba's residential area of Muniki, the army went on the rampage, burning 40 houses.

Damage to culture

Culture may be defined as a complex whole that makes one society different from the other. African society is strongly communal in nature and much is done within the framework of interdependence, common effort and support of one another. But under war conditions the communal element of human co-existence is totally lost, in particular among the displaced people living in tents or on open ground where concern for survival occupies their entire minds. Similar damage is done also to religion which is rooted in the cultural system, respects the norms and values of society and provides understanding of supernatural forces. The war has wrecked these values and has begun to make whole societies to degenerate into a fatalistic attitude.

The deterioration of the status of women

Refugees from the areas of conflict are, unlike those in the regions affected by drought and famine, mostly women and children. This is because men are forcefully drafted into the ranks of guerrilla forces and many of them are killed in the fighting.

Women have a special role in feeding rural households. In most parts of Africa they do more than 80 percent of the farm work. Within households there is a clearcut division of labour: the men are primarily responsible for clearing and land preparation, whereas planting is carried out by all members of the family. Weeding is done exclusively by women, as well as harvesting and crop processing. In the pastoral societies where the basis of subsistence is milk, many people also carry out sporadic cultivation to supplement their food. This is done by women who are also the fuel gatherers and water-fetchers. Vegetable growing, poultry keeping, marketing of food and management of the family's economy is under the women's control. But as the war increases the mobility of people moving from place to place so it increases the burden on women who often must also clear the new land for cultivation and sometimes move on again before harvesting the crop. In refugee camps they feel uprooted and useless and because they have very little, if any, education, they are difficult to train for other jobs.

Violation of human rights

All regimes challenged by armed opponents in Sudan, Ethiopia Somalia, Uganda, Mozambique, Angola and South Africa resort to tough repressive measures against civilians suspected of sympathies for their adversaries. Arbitrary arrests, torture, summary executions,blowing up relief convoys on their way to starving victims of drought, shooting down Red Cross planes, all become a gruesome part of life of the population in the theatre of war.

A respect for basic human and people's rights, including physical safety and food security, is non-existent.This is in particular a serious development in a continent where half of the population has no access to potable water, no money to buy food even if it is available at the markets, no shelter, no opportunity to get even basic education and primary health service. The people have no defence against government-imposed development models. which are destroying their life support system. Popular participation in country development has been replaced by government decrees.

New meaning of security

In Africa the concept of security has acquired a new meaning. In conditions where physical survival is at stake, food security, physical security of an individual and family as well as the whole community is more important than in military terms. Besides, state security concerns become synonymous with the measures aimed at the preservation of power of the ruling elite.

Similarly, the meaning of democracy in Africa has to be linked with the need for a system which would deal with the dissatisfaction with the unjust distribution of national income on the part of underprivileged social groups, a system which would prevent abuses of power and the amassing of wealth by members of the governing regime and their supporters. There is also a need to transform national security to regional security and to provide conditions under which man's capacity to create sustainable societies and to ensure that natural life and support systems will become secure and permanent would receive the government support it deserves.

THE ROLE OF WESTERN EUROPE

The European economic interests in Ethiopia, Somalia and Sudan had never reached the kind of proportions which would call for a protection of these interests by political or military intervention. The European indifference to the political turmoil in the Horn and the decline in the Horn's strategic importance for the super-powers had made the conflicts truly "forgotten wars". It was only after a succession of droughts which produced starvation of hundreds of thousands of people in Ethiopia and Somalia that West European countries responded by a massive humanitarian aid to victims of famine. This also made the Western public aware of the extent of armed conflicts in the region. The food convoys were attacked, Red Cross planes shot down and routes to the refugee camps blocked. In defiance of the international relief efforts, hunger has been used as a weapon by all combating parties. For example, in Tigray province in Ethiopia, where the food shortage is most serious, the assistance through government channels is limited to Mekelle and some surrounding centres, reaching about 700,000 people, which is less than half of the number of starving people living in the areas controlled by the TPLF.

So far, there had been very few signs that Western Europe countries were prepared to become involved in peace initiatives in the Horn. When the British Foreign Minister Sir Geoffrey Howe visited Sudan and Ethiopia in September 1988, all he said was that "The question of peace in the region is a matter of general concern" and added "we shall do what we can to promote that and we ask the Ethiopian Government to do the same". More encouraging had been the recent US offer to mediate between Khartoum and SPLA. It was made by the US Assistant Secretary of State for Africa, Dr. Chester Crocker, at a press conference in Washington on January 27, 1989. He revealed that for quite some time the US had been discussing with the Soviet Union the ways and means of resolving the conflicts in Sudan and the Horn.

THE ROLE OF THE OAU

The African framework which had been used in the past for peaceful settlement of disputes, and which excludes foreign intervention into the peace process, proved to be too fragile to withstand the pressures of violent military coups which swept the continent and of the economic recession which followed. Thus it was the ex-colonial power, the United Kingdom, rather than the OAU which brought about the independence of Zimbabwe, and it was the US and the Soviet Union rather than the OAU Liberation Committee which were instrumental in forging the peaceful settlement in Angola and agreement on the independence of Namibia.

The OAU's insistence on "non-intervention in the internal affairs of States" embodied in Article III of the OAU Charter technically disqualified the OAU from dealing with the conflicts in the Sudan, Ethiopia and Somalia regarded by the Government concerned as "their domestic affair". However, this position is no longer tenable. The OAU and UN principle of non-intervention was discarded when the system of "apartheid" was recognized to be the legitimate concern of all members of the international community of States. By the same analogy, the misery and suffering of the people in the Horn caused by the ecological catastrophe, the effects of which had been multiplied by the armed conflicts, can no longer be regarded as an "internal matter". Unfortunately, there is nothing the OAU can do about alleviating the suffering of the people in the Horn.

THE NEED FOR A REGIONAL APPROACH TO THE RESTORATION OF PEACE, STABILITY AND SECURITY IN THE HORN

The conventional peace negotiations between delegations of the conflicting parties cannot succeed unless they include the participation of leaders of local communities, and unless the search for peace is linked with the search for total security, i.e including environmental security. A regional solution of the conflicts in the Horn presupposes also the participation of Egypt, because of its historical links with Sudan, and of Western European countries, because of their pre-independence ties and economic potential. Any peace solution in the region must be backed by the super-powers. After all, the wars in the Horn are fought with Soviet and US weapons.

The elimination of sources of insecurity and threats to survival of the communities in the region may prove to be a crucial test of the viability of existing political entities as States which have tried to impose a central rule over a multi-ethnic society.

THE DIFFICULT TASK OF POST-WAR RECONSTRUCTION

The cessation of armed hostilities is, of course, only a first step towards the normalization of conditions for life of the people in the war areas.

Rebuilding of destroyed roads, water supplies, reclaiming of land to agricultural use, rebuilding farms and homes and providing basic health and educational services will be a tremendous task. Equally difficult will be the restoration of confidence on the part of the people, notably on the part of the returning refugees into their local and central governments and their law and order enforcement forces which together with the rebel armies were responsible for the devastation of their lives and environment.

A rehabilitation will require a totally new development approach and strategies. These have to be worked out in the closest possible co-operation with the people in the rural areas rather than with the governments only. While the economic recovery, given the availability of funds and expert aid, might be achieved within a reasonable period of time, perhaps five to ten years,the war scars inflicted on the minds of the people might take a generation to heal.

NOTES

1. Tony Hodges, *Angola to the 1990s*, The Economist Intelligence Unit, London, 1987.
2. *Africa Recovery*, vol. 2., No.2, June 1988."Mozambique emergency plan highlights rehabilitation."
3. John Sweeny,"Revenge of the Tall Men", *The Observer*, London, September 4, 1988.
4. The recent fighting in Somalia is well described by Graham Hancock in "Rebels carve a swath of death accross Somalia", *The Sunday Times*, London, September 11, 1988.
5. A major contribution to the understanding of the ruthless conflicts in Ethiopia has been made by nine authors in a recent book,The Long Struggle of Eritrea For Independence and Constructive Peace, edited by Lionel Cliffe and Basil Davidson. It was published by Spokesman, Bertrand Russel House, Gamble Street, Nottingham NG 7 4 ET, England.
6. *BBC World Service*, November 8, 1988.
7. *War Wounds*, see page 18.

Ecological Stress and Political Conflict in Africa: The Case of Ethiopia

Bekure W. Semait

INTRODUCTION

No detailed discussion of political conflicts or ecological conditions is attempted in this paper. Only a brief outline of political conflicts in the three major problem areas, namely Eritrea, Ogaden and Tigray, is given. Major climatic characteristics of the three areas are briefly described to see if there is coincidence between political conflict and ecological stress in space. Not attempt is made to establish cause and effect relationship between the two phenomena. The paper then discusses, again briefly, the resettlement programs in Ethiopia since these are the direct consequences of ecological stress.

AREAS AND CHARACTERISTICS OF MAJOR POLITICAL CONFLICTS

The major areas of political conflict in present-day Ethiopia are three. Two of these, Eritrea and Tigray, are in the northern part of the country; Eritrea bordering the Red Sea and Tigray immediately south of Eritrea. the third area, the Ogaden, is in south-eastern Ethiopia. bordering the Republic of Somalia. These locations are significant from the point of view of the conflict. The conflicts in Eritrea and the Ogaden are inspired by intentions to dismember Ethiopia. What inspires the rebellion in Tigray is stated rather vaguely.

Political conflict in Eritrea

The administrative region that is now Eritrea was known as Bahr Medir and was administrated by a Bahir Negash, viceroy of the Emperor of Ethiopia, until it was occupied by Italy in a process of colonialization lasting from 1882 to 1890.

The process of colonialization started at Assab where an Italian private company had bought a trading post in 1869. The Italian government rebought Assab, and in 1882, it was formally declared an Italian colony. Three years later (1885) Italy occupied Massawa. This was followed by a series of advances onto the highland until the formation of the colony was completed and the area rebaptized Eritrea on January 1, 1890. The Italian occupation was facilitated by the sudden death of Yohannes IV, the severe famine, the cholera epidemic and the cattle disease—all of which occurred in 1889. The resistance offered by the Ethiopian army against the Italian advance, though at times very effective (e.g. Battle of Dogali in 1887) could not be sustained because of defacilitating natural occurrences.

Italy used Eritrea to launch her invasion on the rest of Ethiopia in 1935. As a result of the new conquest, Italy administrated the whole of Ethiopia between 1936 and 1941. Thus Eritrea was occupied by Italy until 1941, the year she was defeated by British and Ethiopian resistance forces. Eritrea remained under British administration for a further 10 years. In 1952 she was federated with Ethiopia. The federal status was abrogated by the Federal Parliament in 1962 thereby making Eritrea one of the 14 provinces of Ethiopia.

Towards the end of the 1950s some Eritreans living in neighbouring countries formed a liberation movement. Since then, a number of movements have appeared, disappeared, made larger unions as well as disunited. Not so infrequently, they have waged bitter political and military struggles against each other.

In the first half of the 1960s the military wings they created started to cause trouble, then to harass and later on to seriously challenge the army. To do this they found natural fortresses in the semi-arid lowlands where nomadism is the main form of life and in the extremely hilly areas that complicate the movements of a regular army.

Their activities intensified and declined in response to military, political, etc. situations in the rest of the country, and to the international climate relevant to them. Thus in 1977–78, when the country was engaged in resisting Somali aggression in the south-east, they intensified their activities and occupied a large portion of the highland. The government was compelled not only to use the army already stationed in the province but also to declare a national mobilization. As a result the secessionists were pushed to the peripheral lowlands and extremely hilly areas, but not put to rout.

After about 10 years, in 1987–88, the secessionists intensified again their activities. As a result, the state council of PDRE issued a special Decree on Emergency Situation in Eritrea on May, 1988. The special decree is now in force.

Political conflict in the Ogaden

The Ogaden is a region in south-eastern Ethiopia bordering the Republic of Somalia and mainly inhabited by a Somali-speaking population. On the basis of language affinity and the idea of "Greater Somalia" the Republic of Somalia claims the Ogaden. In fact it is only one of the three areas outside the Republic that is claimed. The other two are a part of the Republic of Djibouti and the North Eastern District (NFD) of Kenya.

The idea of Greater Somalia was first proposed in April 1946 by Mr Bevin, Britain's Foreign Secretary to the Four Power Commission (Britain, France, the Soviet Union and the United States) which at the end of the Second World War had to decide on what to do with the ex-Italian Somaliland. Mr Bevin proposed that the then British Somaliland, Italian Somaliland and the adjacent areas of Ethiopia which by the Ethio–British Agreement of December 19, 1944 had been put under British military administration temporarily as Ethiopia's contribution to the war effort, be placed under British trusteeship for Greater Somalia. Evidently, the proposal was rejected immediately by Ethiopia.

The idea of Greater Somalia continues to be entertained enthusiastically by the Republic of Somalia since her accession to independence in June–July 1960. This has put the two countries on a constant state of being at war. In addition to the tension on their common border and a number of minor clashes, the two countries fought two major wars.

The first major war was in 1963–64 when the Somali forces invaded across the border into Harerghe. The Ethiopian army could repulse the attack. Further fighting was avoided by the intervention of the O.A.U., which asked the two countries to cease fighting and withdraw their respective army away from their common boundary.

The second and more serious invasion by the Somali army was launched in 1976–77. Ethiopia was in the midst of her revolution and consequently not well prepared to defend herself. Somali troops could overrun a considerable portion of eastern Ethiopia; very much larger than the area she claims. Both of these wars caused considerable loss in material and human resources.

Thus, while the conflict in Eritrea is local, at least in shape, the conflict in the Ogaden is international both in shape and content.

In May 1988, Ethiopia and Somalia concluded an agreement to ease the tension along their common boundary. Among other issues such as ceasing aggressive propaganda against each other and exchanging prisoners of the last war, they agreed to withdraw their troops 15 km from their common boundary. It is hoped that this first step could be followed by further discussions and agreements on the basic causes for their poor relationships.

Political conflict in Tigray

As stated earlier, Tigray is the area just south of Eritrea in northern Ethiopia. It was in Tigray that the Axumite civilization and Empire had their origin. Later on the seat of the Ethiopian government gradually shifted south until it settled in Shewa, making Tigray geographically quasi-peripheral.

Conflicts between Tigrayan chiefs and their counterparts in the rest of Ethiopia for political supremacy were not uncommon in the past. Thus the relationship between Menelik in the south and Yohannes IV in the north was not very harmonious. At the death of Yohannes in 1889, Mengesha's accession to the throne was frustrated by Menelik.

In 1941–43 there was the Weyane rebellion in Tigray. It was suppressed through the use of the army by the central government.

The present conflict in Tigray is caused by the "Tigrean Peoples' Liberation Front (TPLF)". The TPLF was established in 1975 amidst the Ethiopian Revolution as a number of other political groups were. Its causes and goals are rather vague to this author. Statements like "The Tigrean people are called on to take charge of their own lives by challenging the root causes of their suffering and exploitation; imperialism, bureaucratic capitalism and feudalism, and by struggling for national determination, the right to decide their political status by their own free will" are accompanied by other statements such as "we struggle for the liberation of the whole of Ethiopia and we do not have secessionist aims".

Whatever the driving force of the struggle, the TPLF in 1988 made normal life in the region extremely difficult and the government was compelled to mobilize the army against the TPLF and to issue the Special Decree Emergency Situation in Tigray on May 15, 1988, at the same time as it did in Eritrea.

We have briefly surveyed the political conflicts in the Eritrea, the Ogaden, and Tigray regions of Ethiopia. Equally briefly, we shall survey the ecological situation in the three regions for the purpose of finding spatial coincidence between the conflicts and specific ecological characteristics.

ECOLOGICAL STRESS

We are all aware that only aspects of climate do not constitute ecology. In spite of that, our discussions focus on aspects of climate for two reasons: (1) the aspects of climate that we are going to consider briefly are the most important elements of ecological conditions in the three regions of Eritrea, the Ogaden, and Tigray; (2) they are also the most measureable and available of ecological elements. The aspects of climate we shall consider are distribution of annual rainfall, seasonality of rainfall, moisture balance, and reliability of rainfall in Ethiopia in general and in the three areas that concern us in particular. Then the human aspects of ecology are mentioned even more briefly than the aspects of climate.

Four sketch maps are drawn to facilitate our discussion of the aspects of climate mentioned above. At the same time, the location of the three areas of political conflict is indicated.

Distribution of annual rainfall

On map I is shown the distribution of annual rainfall over the country using isohyets. As can be observed, the heaviest annual rainfall, more than 2,000 mm, occurs in the south-western part of Ethiopia. From here, rainfall decreases more or less progressively, in all directions.

Thus, in the NNE direction rainfall decreases to less than 100 mm in Eritrea. The largest portion of Eritrea receives less than 200 mm of rainfall. Tigray is slightly better; however, most of it receives less than 600 mm of annual rainfall. Similarly the annual rainfall decreases in the south-east direction to less than 200 mm in the Ogaden. More than 90% of the Ogaden receives less than 300 mm of annual rainfall.

Hence, it is clear that the three areas of political conflict are located in zones of rainfall scarcity.

Seasonability of the rainfall

As shown on map II, the duration as well as the period during which rain occurs vary considerably. In a zone extending from south-eastern Ethiopia to the highlands of Harerghe rain occurs most of the year, and in a considerable amount. The duration of the rainfall decreases in the NNE direction and on the Red Sea coast, changes from summer concentration to winter concentration. In the south-east the Ogaden has fewer rainy days scattered in two seasons—spring and autumn. Thus, Eritrea, Tigray and the Ogaden are characterized by scanty rainfall and short rainy seasons.

Moisture balance

What determines ecological conditions in an area is not simply the volume of rainfall. The amount of moisture available in the soil, which is not purely a function of precipitation, influences more significantly the interaction between the natural and human elements of ecology. Soil moisture is largely the result of the interplay between precipitation and temperature.

Using the De Martone Index we have sketched map III. Most of Eritrea and the Ogaden fall in the "arid" classification. Tigray is semi-arid and slightly wet. Therefore, the three regions under consideration do not possess the moisture requirements to support large plant and animal population.

Reliability of the rainfall

Whether or not rainfall begins and ends in a regular and predictable manner with more or less the same annual amount, is an extremely important factor with regard to the human use of moisture or water. Such dependability can be indicated by the coefficient of annual rainfall variation as shown on map IV.

South-western Ethiopia has more dependable rainfall than the rest of the country. Here the coefficient or variation is less than 10%. The three areas that interest us here have the least dependable rainfall. The whole of Eritrea and the Ogaden are located in the zone having coefficient of variation between 40 and 90%. Most of Tigray lies in the zone having a coefficient of variation of 31 to 40%.

Thus, Eritrea, the Ogaden and Tigray have an extremely fragile natural ecological constituent.

The state of some other ecological elements

An important ecological element is plant life, which is highly dependent on climate, among other things. Since precipitation in the three areas is meagre, vegetation is poor. Where it is not cleared it is characterized by scanty thorn bushes that are easily disturbed by slight animal and human intervention.

Another environmental problem in the three areas is the recurrent invasion of desert locusts. The fact that locust is still hazardous has been evidenced by the recent plague of locust in most parts of Africa, including northern Ethiopia bordering the African desert zone in the second half of 1988.

Also, demographic factors and its concomitants acting on the fragile environment disturb the ecosystem. Population growth is progressively increasing due to access, even though extremely limited, to medical facilities. This has been increasing the population pressure on the limited resources. Land that has not been considered appropriate for cultivation and grazing is now cultivated and grazed more and more. The systems of land use such as fallowing and rotational grazing have become impractical in many areas.

Cause and effect relationships

We can conclude from the foregoing that there is a spatial coincidence between ecological stress and political conflict in the three areas. What has not come out and cannot be done conclusively here is whether there is cause and effect relationship between the two. We can only make hypothetical statements and cite limited causes of cause and effect relationships.

The Ethiopian lowlands in general and the peripheral lowlands in particular have been avoided by settled agriculturalists. Similarly, with the few exceptions of ports and road or railway junctions no important urban centres developed in these areas. They have been the domain of effective central government administration. As a result, the central government had been unable to protect the people from various forms of external influence at all times.

The external influences were of two types. One of the forms was direct occupation by foreign powers as in the case of Eritrea. The second form of influence was social and ideological as in the case of the rest of the lowlands. It can be assumed that this state of affairs inculcates in the people of these regions a sort of "independent" existence.

The mode of life in the lowlands which was a natural response to the prevailing ecological and socio-technological conditions has been used, though a pure pretext, by foreign powers to advance certain ideas and even to launch military attacks on Ethiopia. As indicated earlier, Mr Bevin's idea of Greater Somalia was advanced under the guise of enabling the nomads to lead their frugal existence with the least hindrance. The Walwal incident between Italian and Ethiopian forces in 1934 was for the purpose of occupying the water wells in the Ogaden close to the border with the ex-British Somaliland. This incident was one of the justifications for the Italian invasion of Ethiopia in 1935.

For mote than a century the regions of Eritrea, Tigray, Wello as well as south-eastern Ethiopia have been experiencing frequent famine years. Drought, an ecological factor, is one of the causes of famine. We may deduce that famine disasters make the people in these areas extremely sensitive and rebellious, attributing their ordeal to government negligence. This is fully exploited by external enemies of the country.

Two important points may be raised at this juncture. First, it is hard to attribute political conflicts to one single cause such as, in this case, ecological stress. Second, even when there are indication that ecological stress contributes to political conflicts, research covering more space and time than this paper has done is needed to determine the degree of its influence.

More directly, ecological stress has caused the launching of resettlement of people from stress areas to more productive ones in the west and south-west of the country. We shall briefly consider this program.

RESETTLEMENT

Famine has recurred periodically in Ethiopia, more particularly in the northern part, over a long period of time. Its basic causes have been drought, pest, hail-storms and cattle disease. The Ethiopian population, ill equipped scientifically, technologically and socio-politically, suffered considerably from the recurrent famine. One response to famine has been spontaneous and sporadic resettlement by the hard struck people to areas that do not suffer the disaster so frequently. Thus, resettlement of people in search of real or assumed better opportunities has been going on for a long time.

The process and volume of resettlement

Planned resettlement started in the late 1950's when population pressure in some parts of highland Ethiopia was considered a problem. Later, in the 1960's, a U.S. Aid mission report supported the idea of resettling people and World Bank assistance was secured to implement settlement projects. However, only some 20,000 families have been resettled before the 1974 Ethiopian revolution.

Planned resettlement received an impetus after the revolution. The famines of 1974/75 and 1984/85 provided the urge and determination to undertake large-scale resettlement programs. Its implementation was enhanced by (1) the land reform proclamation of 1975 according to which rural land became public property, (2) the creation of a unique party—the Workers Party of Ethiopia (WPE) and (3) the establishment and participation of public organizations such as Revolutionary Ethiopian Youth Association (REYA), and Ethiopian Teachers Association. The responsibility of resettling people was given to the Relief and Rehabilitation Commission (RRC) between 1974 and 1976, to the Settlement Authority between 1976 and 1979 and to a strengthened RRC amalgameting the Settlement Authority and the Awash Valley Authority since 1979. Hence, the resettlements shown in Table I were established between 1975 and 1981. The location of the villages and the origin of settlers by administrative region are shown in Table II.

The total number of resettlers was about 120,000. In addition it is estimated that between 1978 and 1983, some 52,000 people were resettled on the Arsi side of the Wabe River Valley.

The most important undertaking was made in 1985 in response to the severe famine that occurred in the northern part of the country in 1984/85. Altogether, 587,785 people were resettled in the western and south-western parts of the country. Of these 62.4% were from Wello, 18.4% from Shewa, 15.3% from Tigray, and 3.9% from Gonder and Gojam. However, it should be noted that Gojam received 17.2% of the resettlers at the same time.

In 1986/87, resettlement was halted for about a year for the purpose of evaluating past activities and planning future actions. Resettlement was resumed in November 1987 and continued more cautiously until May 1988. During this last phase, some 10,000 persons, most of them from Wello, were resettled in western Gojam.

Post revolution resettlement differs from the traditional one in at least four respects. (1) It is government initiated and planned. (2) It is on a relatively large scale.(3) It involves movement over long distance. (4) Assistance is provided until the resettlers are able to support themselves.

Issues regarding resettlement

As indicated earlier, the resettlement is implemented by the government using its party, administrative, and public organizational structures. The areas of resettlement, especially those involving movement over long distances, are areas with low population density or not habited effectively so far. At the same time a lot publicity is done to convince people on the receiving end that facilitating the resettlement scheme is a national responsibility and necessary response to fellow country men in distress. Consequently, the population in the receiving areas participates in building shelters, providing household utensils and draught-animals to resettlers. This has reduced conflict between new and old settlers to almost nothing.

The few incidents encountered are no more than common misunderstandings that take place in a community. These are issues related to the use of grazing land and water, and the collection of honey from bee hives in forests and bearing no property identifications.

However, the resettlement schemes draw negative comments from the western media. The western media carries statements alleging that people are resettled by force. It also alleges that the resettlement is effected with disregard to the interest of the already existing population.

The government consistently states that the resettlement is voluntary and no coercion is used on people to resettle. As regards the local population in the receiving areas, every necessary care is made to respect their interest. But in most instances resettlement is done in uninhabited areas. Hence the allegation by the western media is simply described as emanating from the desire to discredit the revolution.

Another opposition group to the resettlement scheme is the Ethiopian refugees abroad. They feed the international media with all sorts of information. Many of these refugees are grouping themselves in a number of political organizations on the basis of regionalism or tribalism. It is quite possible that such groups, in the future, will try to create political conflict in the resettlement areas by playing the old residents against the resettlers.

The following can be concluded from the discussion so far. There is spatial coincidence between ecological stress and political conflict in the areas that we have considered. A precise cause and effect relationship cannot be established without carrying our further research. The resettlement scheme which has direct relationship to ecological stress has been briefly discussed. Right now, it is a political issue only in the international media. However, care has to be taken so that it may not be used to stir political conflict between the old and new inhabitants.

REFERENCES

Bekure, W.S. 1978: *Climatic Characteristics and Life Patterns in the Semi-Arid Areas of Ethiopia*, (unpublished).

Clarke, J. 1986: *Resettlement and Rehabilitation: Ethiopia's Campaign Against Famine*, London.

Clay, J.W. & Holocomb, B.K. 1986: *Politics and the Ethiopian Famine 1984-85*, New Jersey.

Daniel, Gamachu 1988: *Environment and Development in Ethiopia*, International Institute for Relief and Development, Switzerland.

Daniel, Gamachu 1985: "Peripheral Ethiopia: A look at the Marginal Zones of the country" `In *Regional Planning and Development in Ethiopia, IDR and IREUS*. Addis Ababa. pp. 81–91.

Daniel, Gamachu 1977: *Aspects of Climate and Water Budget in Ethiopia*. Addis Ababa University Press.

Drysdale, J. 1964: *The Somali Dispute*, London.

Eshetu, C. & Teshome, M. 1984: *Land Settlement in Ethiopia: A review of Developments*, A.A.U. Unpublished.

Gebru, Tareke 1977: *Rural Protest in Ethiopia 1941–1970, A Study of Three Rebellions*, Ph.D. dissertation, Syracuse University.

Hailu, W. Emanuel 1964: "Concession Agriculture in Eritrea" *Ethiopian Geographical Journal*, vol. II, No. 1, pp 35–44.

Jones, A.H.M. and Monroe, E.A. 1974: *History of Ethiopia*. London.

Kebede, Tato 1964: "Rainfall in Ethiopia" *Ethiopian Geographical Journal*, vol II, No. 2, pp 28–36.

Mesfin, W.M. 1964: *The Background of the Ethio–Somalia Boundary Dispute*. Addis Ababa.

Mesfin, W.M. 1964: "The Awash Valley-Trends and Prospects" *Ethiopian Geographical Journal*, Vol II, No. 1, 1964, pp. 18-27.

Mesfin, W.M. 1977: *Somalia: The Problem Child of Africa*. Addis Ababa.

Pankhurst, A. 1986: *A Report on a Study Tour of Settlement Schemes in Wollega*. Manchester (unpublished).

Pankhurst, A. 1988: *Responses to Resettlement: Household, Marriage and Divorce*, Xth International Conference of Ethiopian Studies, Paris (unpublished).

Pankhurst, A. 1988: *The Administration of Resettlement in Ethiopia since the Revolution*. Oxford (unpublished).

Peberty, Max 1985: *Ethiopia's Untold Story*. RST, London.

Petrides, S.P. 1983: *The Boundary Question Between Ethiopia and Somalia*. New Dehli.

P.M.A.C., 1975: *Public Ownership of Rural Lands Proclamation*, nr 31, Addis Ababa.

Rubensson, S. 1971: *The Survival of Ethiopian Independence*, A.A.U. Press.

Shaccehi, A. 1985: *Ethiopia Under Mussolini*, London.

Shumet, S. 1984:*The Genesis of the Difference in the Secessionist Movement 1960–1970*. (unpublished).

Suzuki, H. 1967: "Some Aspects of Ethiopian Climates" *Ethiopian Geographical Journal*, vol. V, No. 2, pp 19–22.

Tekeste, N. 1987: *Italian Colonialism in Eritrea, 1882–1941*. Uppsala.

TPLF; People's Democratic Program of the Tigray People's Liberation Front, Second Congress 1983.

Ullendorff, E. 1973: *The Ethiopians: An Introduction to Country and People*. Oxford.

UNDP/RRC-Eth/81/001, The Nomadic Areas of Ethiopia, Study Report Part III, Socio-economic Aspects.

TABLES AND MAPS

Table I. *Establishment of resettlement villages*

Year established	No. of villages
1975	5
1976	14
1977	10
1978	11
1979	10
1980	32
1981	5
Total	85

Source: Eshetu & Teshome

Table II. *Origins and location of resettlers*

Location and No. of villages		Origin of Resettlers	
Region	No. of villages	Local	Outside
Wellega	25	6	19
Shewa	20	19	1
Bale	12	-	12
Harerghe	8	-	8
Gonder	4	4	-
Sidamo	4	4	-
Wello	4	4	-
Gamo Goffa	3	3	-
Kefa	3	3	-
Eritrea	1	1	-
Illubabor	1	1	-
Total	85	45	40

Source: Eshetu & Teshome

Map I. *Average annual rainfall*
Isohyets, mm

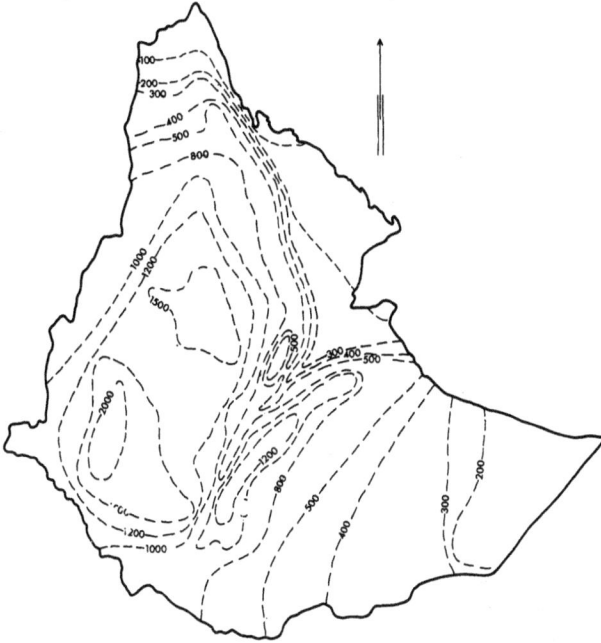

Map II. *Seasonality of rainfall*

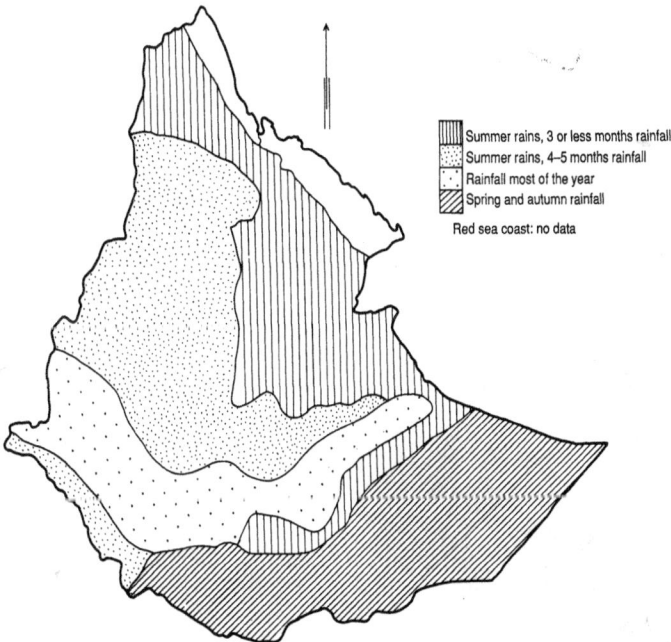

Summer rains, 3 or less months rainfall
Summer rains, 4–5 months rainfall
Rainfall most of the year
Spring and autumn rainfall

Red sea coast: no data

Map III. *Moisture balance*
de Martone index

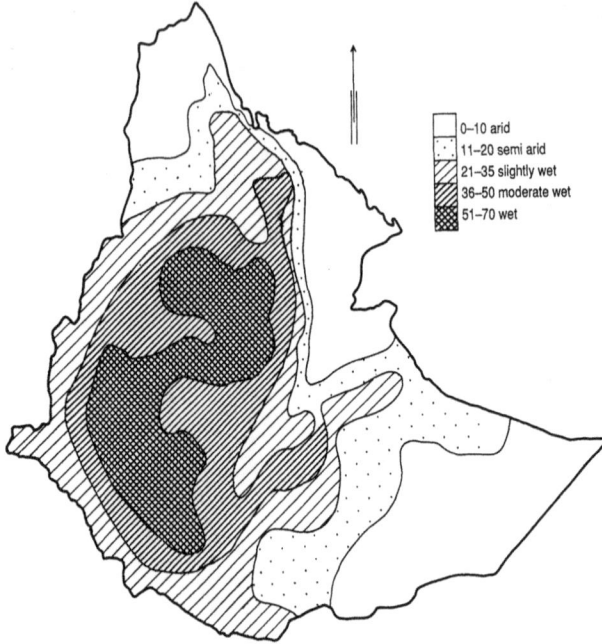

0–10 arid
11–20 semi arid
21–35 slightly wet
36–50 moderate wet
51–70 wet

Map IV. *Coefficient of variation*
Percentage

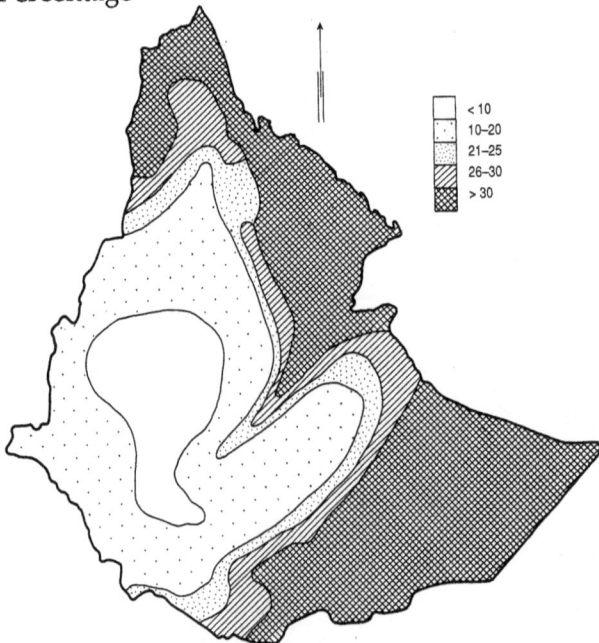

< 10
10–20
21–25
26–30
> 30

Environmental Degradation as a Consequence of Socio-Political Conflict in Eastern Mara Region, Tanzania

Carl Christiansson and Eva Tobisson

INTRODUCTION

This essay is an attempt to explain observed land degradation in Eastern Mara Region in Tanzania not only as an effect of general increase in population but rather as a consequence of the particular socio-economic cultural and political conditions in this remotely situated and rather isolated part of Tanzania.

The results are preliminary and the views put forward here will be further elaborated by the authors in a forthcoming article which will take into consideration also conditions on the Kenyan side of the national border which forms the northern boundary of Mara Region.

In October 1986 one of the authors of this paper (Dr C. Christiansson) carried out a study for SIDA to assess the status of the environment in four districts in Mara and Mwanza regions, northern Tanzania. The aim was to investigate possible impact on land, water and vegetation of a SIDA-sponsored water supply project (Christiansson and Kikula, 1986). Particular efforts were made to assess the situation regarding water for humans and livestock. The state of grazing, timber and fuelwood resources and present trends regarding degradation of the land were also observed. One of the selected districts was Serengeti, which together with eastern Tarime District is the area in focus in this paper.

During the course of the study it became clear that in this part of the country unstable socio-political conditions Prevail which do not only negatively affect the water development programme but also add to the general degradation of the natural resources in eastern Mara Region. To make Possible an analysis of the causes behind the socio-political situation Dr E. Tobisson, Social Anthropologist with extensive experience of Mara Region joined as co-author of this paper.

LOCATION OF THE STUDY AREA

The location in East Africa of the area dealt with in this paper, Serengeti District and the eastern reaches of Tarime District in Tanzania, is shown in Map I. The area is bounded by Masai Mara National Park in Kenya in the north–east and by Serengeti National Park in Tanzania on its southern and eastern sides.

The parks have extensive influence on this area, which is more or less their direct extension as evidenced by the large numbers of game migrating through the districts. Besides, the Serengeti Park now acts as a buffer zone separating several agro-economic systems.

The pastoral Masai who used to cross the park areas, either for raiding or trade purposes, now remain east and north of the parks. The land along Lake Victoria and major rivers are inhabited by Bantu-speaking and Nilotic groups (e.g. Suba. Kwaya, Jita and the Nilotic Luo) who to varying extents supplement agriculture and livestock-keeping with fishing. The number of Bantu-speaking groups practicing an agropastoral economy are many, e.g. the Sukuma in the south and the Kuria, Zanaki, Ikizu, Ngorimi, Ikoma, Isenyi and Nata. Some of these groups also engage in hunting. The diversification of the economic systems around the Serengeti made it a major cross-road of commerce which fostered friendship and alliances, but from time to time conditions of hostility also arose between various groups.

SOCIAL AND ECONOMIC INSTABILITY

The most obvious sign of today's unrest is the frequent cattle raiding, particularly in Serengeti District, and the losses of human lives associated with the cattle thefts. From January to August 1986, 25,500 cattle (between 10 and 15% of all cattle in Serengeti District were stolen and more than 100 persons lost their lives in these operations. Moreover several large villages were burnt down (Daily News, Oct 1986).

In the same period another 20,000 cattle and thousands of sheep and goats were shifted from their customary grazing grounds to what was regarded as more peaceful areas near Mugumu, the Serengeti District centre, and to neighbouring districts. This situation has contributed to the growth of Mugumu town from around 5,000 inhabitants in 1978 to double that number today. At the same time there is a migration away from Mugumu, as traders, shop-keepers and others who feel insecure even in the township and whose business is negatively affected by the unrest, tend to leave the district. In the vicinity of the district centre refugee. Problems can be observed, highlighted by deforestation and increased pressure on farm lands and food and water resources.

The problem of cattle theft involving the loss of human lives is most severe in areas between Mugumu town and the eastern part of the adjacent Tarime District. The problems are so severe that many district officials express unwillingness to visit these areas and the state of insecurity also causes unwillingness among local people to participate in communal development projects, e.g. digging of wells.

As we shall see below, the area is inhabited by numerous Bantu-speaking groups which exhibit fundamental language and cultural similarities, manifested in e.g. intermarriage. But the different groups also regard each other as potential enemies competing for scarce resources. The raiding which takes place today. with ensuing unstable socio-political conditions and environmental degradation should be looked upon as an extension of these customary relationships in which one ethnic group, the Kuria, has proved to be militarily superior.

PRESENT RESOURCE SITUATION

The region has a general slope from 1,800 and 2,000 m in the highland north and east to 1,200 and 1,300 in the lowland west and south, with dramatic escarpments separating the northern highlands from the flatter lake shores and southern plains. All rivers draining the area flow into Lake Victoria. The climate ranges from sub-humid to semi-arid. The northern highland areas have deep and fertile soils and a well-distributed annual rainfall of 1,250 to 1,500 mm. Rainfall in the lowlands is limited and erratic, the average for thirty years (1931–1960) for stations in this area being 800 mm per year. Given these conditions the availability of water is here restricted and the recharge of the sources used is irregular and uncertain.

The conditions prevailing in the lowlands place important limitations on agricultural production and on the carrying capacity of the pasture areas. While the highland population can harvest two crops per year, the conditions in the lowlands provide options for one cycle only. In earlier times the hazardous conditions in the lowlands periodically resulted in great losses of human as well as animal lives. Tanner (1961), who made a demographic study in the area in the period just before Independence, found that of internal migration in one division, 50% was caused by human or stock deaths while shortages of water or food, scarcity of land or too much vermin, accounted for 25%.

Scarcity of water

Mugumu township, the district centre, was established in 1974 when Serengeti District was created. Before the township was formed, the settlement pattern was dispersed and people relied on scattered water sources.

Today Mugumu suffers from a severe scarcity of water. This has been aggravated by the many refugees now staying in the vicinity of the town.

The severity of the water problem has implied that in 1986 water was being pumped only once a week for a few hours and only to parts of the town. A few years ago three shallow wells in the vicinity of the town were fitted with hand pumps. However, the pumps were soon destroyed by the users and were then removed.

Some low-yielding natural springs around the township serve as an important source of water utilized by many households. At peak periods when the springs are intensively used it takes up to two hours to fill a 20 litre bucket. Due to the low yield of water people have to stand on the floor of the well itself in order to reach the water. This practice negatively affects the quality of the water as does also the habit of letting livestock drink directly from the springs. Calves and small stock trample around the wells thus implying further pollution of the water.

The difficult situation has led to water being sold in the town. The normal price for a bucket of muddy water from one of the springs or from tap water (when available) was 10/- in October 1986.

As a consequence of the precarious water situation, water related diseases are very common in the area, as confirmed by the medical doctor at Mugumu Hospital.

Many of the villages in the rural areas have large populations and water for both human and livestock is critical. People frequently walk 5 km or more to draw water. The main water sources are temporary pools or rivers which are dry for most parts of their courses except in the height of the rainy season. Small wells dug in the sandy river beds provide water where other sources are not available.

Water for livestock is a serious problem during the dry season, particularly as there is competition between livestock and wild game. The problem is most acute during years of low rainfall as in 1986. During such years of extreme drought. wildlife move around everywhere looking for water even on the outskirts of the district centre.

The problem has been further aggravated by the fact that the inhabitants of the densely populated highlands further north, where pastures are limited, bring their livestock to the lowlands for grazing.

Dwindling timber and firewood resources

It is apparent that in many parts of Serengeti District the "original" vegetation was woodland of miombo type which provided suitable habitats for tsetse flies. As late as 1967 the area around Mugumu town was dominated by dense woodlands and Serengeti District used to be an important source of timber. Some areas like Ngoreme in the northwest and Isenye and Nata in the southwest that were settled at an early stage have long been characterized by open landscapes.

Currently, woodland in the district can only be found in areas bordering the Serengeti National Park, in the Ikorongo and Gurumeti River game controlled areas. There are also considerable areas of woodland left between Nyichoka and Nata villages south of Mugumu.

Most of the other parts of the district have had their woodland cover reduced to isolated trees and bushes because of a number of reasons:

Clearing for cultivation and clearing for erradication of tsetse flies are probably the most important reasons for the dwindling of the woodland resources. Grazing and browsing of livestock and burning for improvement of grazing areas have also played their role as well as fires set by hunters. Firewood collection, charcoal burning and lately also brick burning has further depleted the resources. Cutting of poles for building purposes constitutes an increasing problem with the great influx of refugees to the central parts of the district. Another reason directly linked with the insecure conditions is the preference for particularly stout trees for construction of cattle bomas which can withstand raiders.

It is noteworthy that in Serengeti District building using burnt bricks is not only restricted to the District headquarters at Mugumu. Houses constructed of burnt bricks are a common feature in many villages along the Mugumu–Bunda road, for example. Villages so far not engaged in brick-making are now taking up this activity as a communal undertaking, a fact which further depletes the woody vegetation. The area around Mugumu is now almost devoid of woody vegetation, thus the wood source for the brick kilns has progressively been extended outwards, increasing the extent of the treeless areas.

The scarcity of woodfuel is reflected in the prices paid for this commodity. In 1986 in Mugumu town one bundle of firewood (10 pieces: 1 m long, 1.5 inch thick) was 20/- Tsh (0.6 USD), while a bag of charcoal of not more than 40 kg ranged between 250–300/- Tsh (7–8 USD). The tree scarcity is also reflected by the use, in many villages, of cassava sticks for fuel.

Decline in grazing and agricultural potential

According to the 1984 national livestock census Serengeti District accommodated roughly 200,000 cattle, 100,000 goats, 70,000 sheep and 450 donkeys.

The actual grazing pressure should, however, be much higher than the figures show, since livestock brought in for grazing from neighbouring districts are not counted.

Soil erosion due to overgrazing occurs in several villages, especially just north of Mugumu. The development of gullies has been rapid, particularly during the last five years. Many gullies are 3–4 m deep and up to 10 m wide. The explanation expressed by the district administration is that the pressure on the grazing resources has increased as cattle owners move south to safer areas from harassed villages in the north.

There is overgrazing also in the northern areas but many of the typical features of soil erosion are absent due to the relatively flat landscape. The large herds of sheep and goats in these areas contribute to the overgrazing.

Generally, the southern part of the district is not as badly eroded or overgrazed, mainly due to the presence of tsetse-flies which limits the herd sizes. Cases of sleeping sickness in humans have been recorded in the area.

The expansion of tsetse is given as explanation by the district authorities as to why the inhabitants of a village in southern Serengeti District recently abandoned their village. At this village there is a borehole which yields abundant water but still the villagers moved away, leaving behind a school, a dispensary and the water supply.

In 1986 there were 24 cattle dips in the district. Lack of water for the dips was a permanent problem. Four dips situated within reach of ample water were out of use because they were situated between enemy villages.

It was not many years ago that Serengeti District was self-sufficient in food, but this is not the case any more. Glaring evidences of the former agricultural productivity include the empty godowns, e.g. at Nata and Robanda villages.

Reasons for this decline in agricultural productivity can be found both in low rainfall and in the insecure conditions under which the farming population has to produce their crops.

LAND OCCUPANCY AND UTILIZATION

The present inhabitants of Serengeti District and the eastern part of Tarime District can rightly be described as hybrid people. Their history is one of dispersal and amalgamation as communities split up and wandered farther afield to new settlements in the hope of finding better opportunities for agriculture, animal husbandry, fishing and game-hunting.

The Ngorimi, Isenye and Ikoma People moved into the present Serengeti District from the east, following drought and Masai attacks on their settlements. The Nata claim that their ancestors hunted wild game north-east of what is now Mugumu town. Zanaki oral traditions relate that their ancestors moved in from different directions, e.g. from the islands outside Musoma, from south of the Lake and from the Kuria territory further north.

The majority of Kuria clans trace their "origin" to an area beyond Mount Elgon, from where they migrated to their present territory on both sides of the Kenya–Tanzania border. The Kuria claim that most of the present inhabitants of Serengeti District descend from Kuria ancestors who proceeded further south from what is now Tarime District.

The people living in Serengeti and Tarime Districts practice an agro-pastoral economy, meaning that they derive their livelihood from a combination of agricultural production and animal husbandry. Due to environmental differences between lowland and highland, however, the relative emphasis on agriculture and livestock production is bound to vary. Agriculture in Serengeti District is mainly at subsistence level, the staple crops being cassava, maize, sorghum and millet. The crops grown for sale are cotton, simsim and groundnuts. In the densely populated Kuria highlands of Tarime District. maize is grown for the twofold purpose of subsistence and marketing and ploughs are commonly used for land preparation. Other subsistence crops of importance in this area are sorghum, finger millet, sweet potatoes and cassava.

Since agriculture is a risky and unreliable enterprise in the lowlands, most of the Serengeti inhabitants depend on livestock for their subsistence. The situation is different in the densely populated, fertile and well-watered Kuria highlands across the Mara River where people accordingly hold large herds of cattle, although not primarily for subsistence needs. Agricultural production in this area necessitates livestock, since female agricultural labour is secured by the use of livestock as bridewealth (average rate of bridewealth presently ranging between 35 and 40 head with cases of 55 head recently recorded).

The northern part of Serengeti District is well known for its large cattle herds. It is an area of reasonably good dry season pasture due to the availability of water in the perennial Mara River. In the 1950s these areas. particularly those away from the river valley were very sparsely inhabited and represented the target areas of expansion in the district at that time. In the late 1950s there was however a considerable immigration from the north.

Tanner, 1961 described this development:

"Along the Mara River in the Ngoreme and Kiaoata areas the increase attributable to aliens from the north totalled about 15 percent of the total population."

Immigrants move into vacant areas where they are easily observed. This is particularly noticeable with the Kenya and North Mara immigrants who tend to occupy flat valley bottoms where they can use ploughs and tractors and where the local inhabitants do not cultivate. There is thus no ecological clash between these immigrants and their hosts and little likelihood of political repercussion in the near future.

The majority of those which Tanner described as "aliens from the north" were Kuria from Tarime District and from across the border in Kenya.

The highland Kuria depend on pastures in the lowlands beyond the Kuria territory, particularly in the last few decades when the increased surplus from agricultural production, in the absence of alternative means of wealth accumulation, has meant a dramatic expansion of the Kuria herds. The fact that ploughs could be used in the valley bottoms further attracted the interest of Kuria immigrants who had been using ploughs since the early 1940s.

In the south of Mara Region most of the soils are of a heavy nature and the only well-drained soils are found around the hills which in some years support good yields of cassava and maize. Around the villages, the vegetation has always been cleared for fear of both human intruders and animal predators; but all the areas beyond the village outskirts are bush-covered and abound in game. Into these areas there has been a continuous influx of Sukuma from the SW. Tanner, 1961, described the situation in Ushashi around the time of Independence thus:

"Kenya migrants form only 4 % of the total population but the Sukuma who come in search of cotton-growing land have flooded nine out of the chiefdom's 17 parishes so that there has been an increase of 27.6 percent in five years. The overall increase in tax-paying population was 77 percent. This is a much more dangerous situation as they like the same sort of land as the local inhabitants and it is interesting to record that this area has produced more than its fair share of political tension in the period under review"

A sparse population is characteristic for those areas where tsetse-flies occur or where water is absent but the general situation is that settlements are rapidly enroaching on the bush. The situation is, however, not universal, as observed by Tanner, 1961:

> "At Ikoma the population reduction has been so marked that the cultivated land and the land cleared by grazing and firewood collection can no longer keep clear sufficient areas to balance the surrounding tsetse bush with the needs of the mixed farming. Almost in front of the eyes of the onlookers this settlement is moving to extinction away from the massive concentration of population occasioned in the nineteen hundreds by the protection of the German military post against Masai aggressiveness"

The restriction in areas of settlement may partly explain the observations made that in Serengeti District the villages seem to be larger than in many other parts of Tanzania. This concentration of population not only leads to overutilization of the land resources, for agiculture, building materials, and fuel but it also means that the limited natural sources of water cannot meet the demand, particularly not during the dry season.

GAME, LIVESTOCK AND THE EFFECT OF THE ESTABLISHMENT OF SERENGETI NATIONAL PARK

One of the deep marks of environment on the history of this area has been caused by the proximity of abundant game. Game has always played a very important role in the people's day to day life, as evidenced by the common use of animals (e.g. zebra, leopard) as totems in clan identification. Game also played an important role in livestock accumulation by providing meat for the people, thus allowing them to preserve their livestock for other purposes.

The ecological extension of the Serengeti National Park has limited human productivity in two respects. First, because the eastern areas are dry and comparatively infertile, it is hostile to human habitation. As a consequence, the people have tended to contain themselves to pockets near rivers and swamps. The result being that most of these areas, already at an early stage, became overstocked and overcropped. The situation is aggravated by the fact that further expansion to the western parts, which are more fertile, has been sealed off by the presence of cattlekeeping cultivators who will not condone more livestock herds in the area. The northwards movement is inhibited partly by the Mara River and partly by a large southward movement of Kuria in search for good pasture lands. It may be mentioned here that highland Kuria family heads commonly let a wife and her dependent children build a homestead in the lowlands to which stock are brought for grazing. Such a measure implies that the benefits of animal husbandry and agriculture, as the two cornerstones of the Kuria family economy, can be maximized.

PRECONDITIONS OF INTER– AND INTRATRIBAL CONFLICT

Although most of the ethnic groups in the area look upon themselves as people of "similar" stock, manifested in a common history of migration, share cultural values, principles of social organization and language, they also regard each other as potential enemies competing for scarce resources. The same applies to some of the clans forming a single ethnic group, as in the case of the Kuria. Each ethnic group and each clan forming part of such a group claim usufruct rights to a specific territory, based on legends relating how their founding fathers were the first to arrive here and to clear the land. It should be borne in mind that the function of oral traditions is to explain and to legitimize the reality of the present, rather than to present a detailed account of what actually happened in the past. The "past" may thus be reinterpreted, revised or even invented, in order for the past to explain the present better (e.g. Rigby, 1985). An example of this is groups of Kuria who presently move into the northern and north-eastern parts of Serengeti District claiming to make use of their "customary" grazing grounds.

Although most of the ethnic groups inhabiting the area under study exhibit a similar social and political structure, they differ as far as military strength is concerned. The Kuria are the most powerful, having an elaborated indigenous military organization for warfare and defense, based on a firm age-set structure.

THE ROLE OF CATTLE RAIDING IN KURIA SOCIETY

It is a well-known fact that the majority of those undertaking the violent cattle raids into Serengeti District, and who bring their cattle there for grazing, are Kuria agro-pastoralists whose territory extends from northern Serengeti District into the eastern parts of Tarime District and across the border to Kenya. There are several complex reasons why the patrilineal Kuria engage in raiding and why their need for pastures are increasing. Some of the more important reasons will be summarized below (more detailed information and analysis is provided in Tobisson, 1986).

The militarily strong Kuria have a long tradition of cattle raiding as part of warfare between alien Kuria clans and between the Kuria and neighbouring ethnic groups. An important function of raiding is to confirm and to reinforce the vital social and political institutions of clans and age-sets on which Kuria society and culture are based. An age-set is composed of individuals, men and women, who have undergone circumcision together. While the importance of age-set membership for women is reduced after marriage, men retain a strong group identification throughout their lifetime.

While cattle-raiding used to form an integral part of warfare. carried out by men forming an age-group which was senior enough to take on military responsibilities (including defence), the present raiding is to a large extent undertaken by individuals and small groups of younger men without the consent of the elders. Thus, whereas stealing cattle from the enemy in the past was sanctioned by elders who on the strength of tradition got a major share of the catch, much of the cattle obtained today through raiding are appropriated on the basis of non-customary criteria, i.e., they belong to the participants of the raiding party. While traditional warfare and raiding was carried out by use of spears and sticks, the men engaged in cattle raiding today use powerful weapons obtained during military service in the war between Uganda and Tanzania in 1979, in which a large number of young Kuria men participated.

It should be noted, however, that the diminishing military role played by age-sets does not imply that the indigenous kind of warfare no longer prevails. There have been a few major incidents in recent years for which the origins must be traced back to the customarily hostile relationships between some of the Kuria clans. But warfare has to a large extent given way to smallscale raiding, where raiding parties may be composed of age-set members or not.

There are two principal factors supporting the kind and intensity of raiding taking place today. First, the high rate of bridewealth makes it increasingly difficult, particularly for young men, to raise the number of animals needed through a surplus from agriculture, wage employment, etc., making raiding and theft a supplementary means to expand the herd. This link between a high bridewealth and intensified cattle raiding in Kuria society was pointed out already in the 1950s by Ruel, 1959. The same phenomena have been documented for other East African societies (e.g. Winter, 1978). Secondly, the fact that the Kuria as an ethnic group is domiciled in two countries which offer different economic opportunities is no doubt conducive to raiding. Whether raiding in the area under study is carried out by Kuria living in Tanzania or Kenya, the easiness through which the border can be passed implies that the captured animals can be sold in Kenya at a high price.

AGRICULTURAL LABOUR AND CATTLE IN KURIA ECONOMY

A most important factor accounting for the increased cattle-raiding and the need for pastures beyond the Kuria territory is thus that the demand for livestock as bridewealth payments has been on a steady increase accompanying an agricultural development process in highland Kuria settlements which has made female agricultural labour increasingly important.

Although tradition has assigned women the bulk work on a Kuria homestead, their present workload relative to that of men greatly exceeds the customary prescripts. A farm economic survey in a highland Kuria village, covering the two agricultural cycles in 1977–78, revealed that women and men accounted for 85 and 15 percent of the total number of hours devoted to agricultural work by the participating husband/wife units (Tobisson, 1980). The most important reason for the increased workload on women is the widespread use of ploughs to extend the area under cultivation, while leaving subsequent agricultural operations unmechanized.

Another reason for the increase of workload on women is that maize, rather than crops such as coffee or tobacco. has been promoted in the Kuria territory. Maize is grown by the Kuria for the twofold purpose of family food consumption and exchange or marketing, a fact which makes it an excellent crop for so-called *emongo* production (*emongo* signifying "reserve"). Different from fields alloted to wives by the family head (aimed to secure the subsistence requirements of herself and her dependent children), *emongo* fields are to be cultivated jointly by family members for the object of filling a specified homestead granary which can be relied upon should the granaries of wives be emptied before a new harvest is reaped. The produce remaining in the *emongo* accrues to the family head.

The male family head who disposes of *emongo* produce, either he waits until a new harvest has been reaped or he decides to market the maize at an early stage, does so for the object of obtaining livestock, usually as a preparation for marriage (either he wishes to marry a wife of his own or to fulfil his customary obligation to marry a first wife for his son).

In this situation as described and analysed by Tobisson (1986) senior men are clearly at an advantage over sons as far as access to marriage cattle and therefore also female agricultural labour is concerned. They exercise their authority on the strength of tradition.[1] The young men have no other choice but to use the marriage cattle reserved for *their* sons, or to engage themselves in raiding and theft.

The desire for cattle should be understood in view of the fact that the Kuria lack alternative means of wealth accumulation. The local markets have always been short in supply of manufactured consumer goods, agricultural inputs (besides iron-ploughs), modern building materials, etc. The remoteness of the area and the poor and unreliable roads have restricted trade in such goods. Yet, the most important reason why the Kuria have persistently invested their agricultural profit (and savings from wage employment) in livestock is the firm conviction, based on customary ideals, that livestock *is* wealth. The assertion that livestock is analogous with wealth is intimately linked with the firm rule that bridewealth should rightfully be paid in cattle. Thus, without bridewealth there can be no marriages and hence no wives to manage agricultural work.

It should be noted that raiding and harassment of neighbouring clans and ethnic groups are not only means through which the Kuria are able to expand their cattle herds, but it also gives them access to pastures which they need in order to maintain their herds.

CONCLUDING REMARKS

The environmental degradation described in this paper is insignificant compared to the dramatic destruction of natural vegetation and agricultural land that occurs in connection with extensive international or civil wars. However, the deterioration of resources in parts of Mara Region is directly related to a conflict situation with many violent elements and with deep social and political implications.

Apart from the environmental degradation, the conflict situation, which has existed in its present form for many years, has led to hundreds of people being killed and many more have lost their animals or have had their homes destroyed.

It is important to point out that although the violence is sometimes a manifestation of sheer criminality it also has its roots deep in the traditional socio-political and economic system as described earlier in

---

[1] The use of cooperative work-teams, which are customarily designed to ensure subsistence needs, to procure an agriculture surplus for sale or exchange, is one example of Kuria men (especially senior family heads) applying strategies based on "traditions" to secure a large female labour force. The members of such teams are morally obliged to take part. The customary plight of Kuria women as the prime agricultural toilers is reinforced by the own expectations of marital life, likened to the life of a donkey. Other strategies are described in Tobisson, 1986.

this paper. What used to be intertribal fighting or "internal" fighting between members of separate Kuria clans now spreads and today larger, less well-defined groups of the community are directly or indirectly involved. The situation is highly aggravated due to the raiders having relatively easy access to modern firearms and where these are not easily obtainable, home-made guns made of stolen water pipes are used.

The conflict situation hinders development efforts and leads to an uneven intensity in the land use. Some areas, particularly near Mugumu, are heavily over-utilized while those areas considered to be insecure are abandoned and taken over by bush. Due to the hostility between villages a number of cattle dips and water supplies can not be utilized. In some areas this accentuates already existing competition for water between wild game, domestic animals and man.

Deforestation is intense. One of the reasons for this is that cattle "bomas" have to be exceptionally strong to withstand raiders' attacks. Thus as thick trees as possible are needed for construction and this takes a heavy toll on the remaining stands of large trees.

With the socio-political situation as unstable as it is, people are very reluctant to invest labour or money in land. Soil conservation is a very unrealistic thought. And there are few promising expansion areas within or around a district like Serengeti. The Serengeti National Park sets firm limits for expansion in the south and east and the national border in the north is an equally effective barrier. The land in the west is already relatively densely populated and not particularly fertile.

Raiding also has an economic dimension which derives from the fact that the Kuria live on *both* sides of the border between Tanzania and Kenya. The proximity of the border means that many of the stolen cattle are not added to the capital of Kuria households in Tanzania but driven across the border to Kenya where they are sold at a high profit. To what extent Kenya-based Kuria are involved in actions in Tanzania is not known to the authors but one would assume that extensive smuggling of livestock as well as other goods occurs.

The political dimension attributed by the Tanzania Government to the conditions in eastern Mara Region became obvious when, in the mid 1980s, a high ranking army officer was appointed district commissioner in Serengeti District. However by 1986 this move had not yet had any significant effect on the general insecurity in the district and the ongoing resource deterioration.

REFERENCES

Anacleti, A.O. 1977: Serengeti: *Its People and Their Environment.* Tanzania Notes and Records. No 81/82. pp. 23–34.

Christiansson, C. and Kikula, I.S. 1986: *Pre-study of Land, Water and Vegetation in Mara and Mwanza Regions, Tanzania. An assessment of Problems to be considered within the HESAWA-programme.* INFRA AB, Stockholm. 47 P + appendices.

Rigby, P. 1985: *Persistent Pastoralists. Nomadic Societies in Transition.* Zed Books. London

Ruel, M. 1959: *The social Organization of the Kuria. A field-work report.* Mimeo.

Tanner, R.E.S. 1961: *Population Changes 1955–59 in Musoma District, Tanganyika, and their Effect on Land Usage.* East Afr. Agr. For J. Jan. 1961. pp. 164–169.

Tobisson, E. 1980: *Women, Work, Food and Nutrition in Nyamwigura, Mara Region. Tanzania.* Tanzania Food and Nutrition Centre, Report No 548.

Tobisson, E. 1986: *Family Dynamics among the Kuria.Agro-pastoralists in Northern Tanzania.* Gothenburg Studies in Social Anthropology 9, University of Gothenburg, 233 p.

Winter, E. 1978: *Cattle Raiding in East Africa: The Case of the Iraow.* Ethology, 2:1, pp. 53–59.

Map I. *Location of Mara Region, northern Tanzania*
The areas studied are Serengeti District and eastern Tarime District.

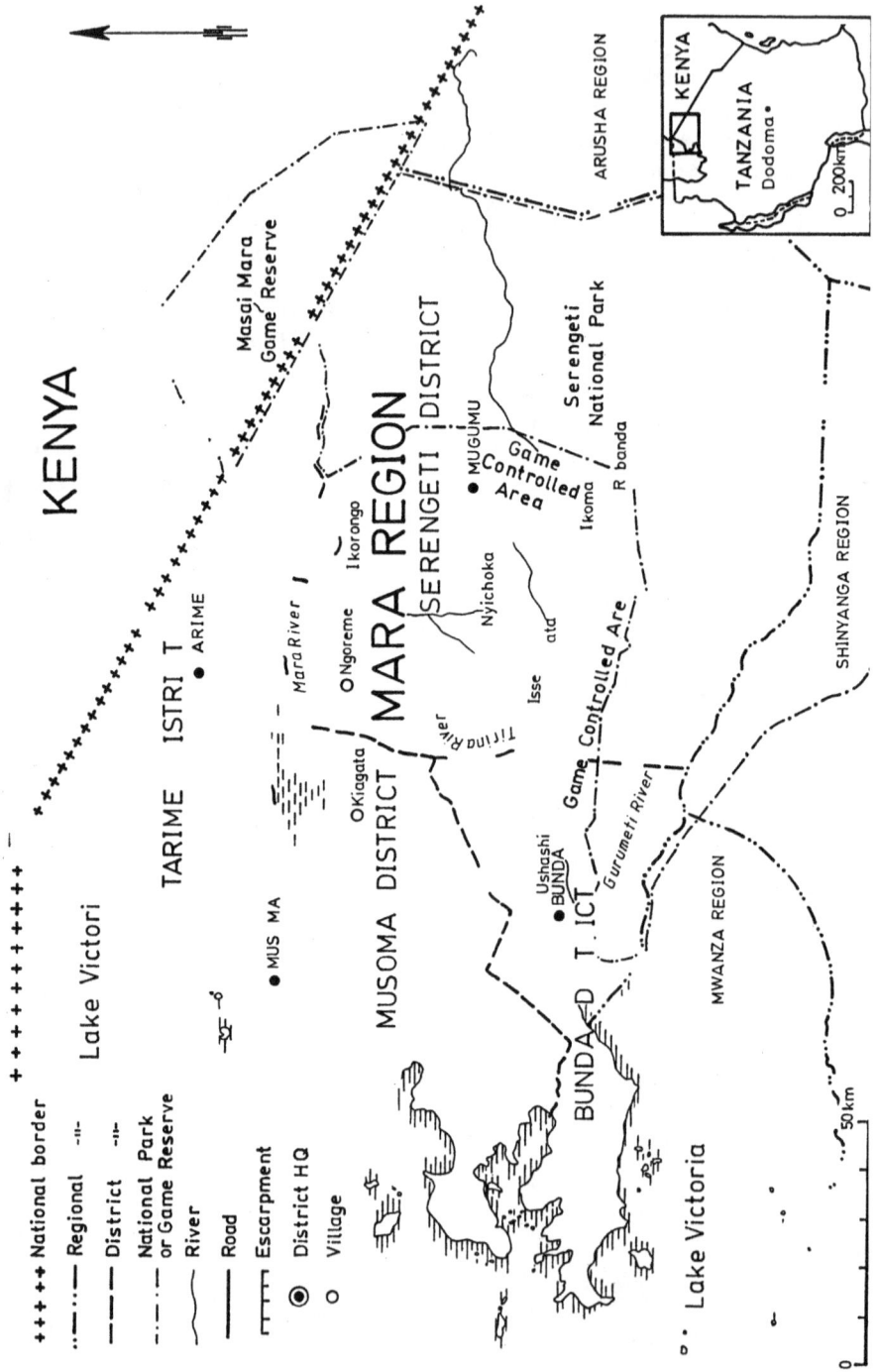

Environment and Security of Dryland Herders in Eastern Africa

Anders Hjort af Ornäs

This paper concerns the living conditions of dryland herders in eastern Africa. My ambition has been to try a local level approach, even down to individual level, in order to see to what extent such an "extended case" approach could raise principal issues for other structural levels; national and regional. I draw upon my experiences in northern Kenya and northeastern Sudan and to some extent also southern Somalia[1]. Living conditions are harsh, life often being a constant fight to establish economic, political, social and even cultural security for the individual. In this paper I wish to concentrate on times of production difficulties, assuming that it is then that we see most clearly interactions between ecological stress and political conflict. In a way the aim of the paper is to offer a community perspective on security, considered in addition to other ideas, be they of a national or international character. The conclusions of the paper are along these lines, and I indicate some attitudes and approaches which I feel are of the utmost importance before we even start using notions such as sustainable development in the arid and semi-arid tracts of eastern Africa.

THE PROBLEM: INDIVIDUAL SECURITY IN TIMES OF ECOLOGICAL STRESS AND POLITICAL CONFLICT

The notion of security is a key word in disciplines such as political science or peace research. We operate significantly with the state as the smallest unit for analysis, possibly with a footnote about regional differences within a country. Similarly, "room for manoeuvre" or "total reality room" are expressions more from the anthropological field. Here we prefer to deal with individuals, or rather households, as the unit of analysis. An outcome of the former approach for the study of eastern African semi-arid tracts is a preoccupation with strategic, even military, considerations of the relations between neighbouring countries. This is of course important for the understanding of living

---

[1] The empirical material comes from the situation for dryland herders in Eastern Africa. It emanates largely from joint work between the author and Gudrun Dahl.

conditions in the rural areas. The efforts within the Intergovernmental Authority on Drought and Development, IGADD, provide a highly relevant example. This is a secretariat at foreign minister level to deal jointly with arid and semi-arid land problems among the member countries Djibouti, Ethiopia, Kenya, Somalia, Sudan and Uganda. Clearly the organization can be seen as a political effort to counter conflict over these often disputed areas and concentrate instead on combating ecological stress.

Looking into options available in such an approach one possibility would be to deal with the substance matter, interrelations between ecological stress and political conflict. This does not, however, lead us to the actual living conditions of today. A more fruitful method of understanding these is to depart from a local perspective and see how the problems appear to those actually living within the development process. The notion of security that I wish to employ will be that of micro/macro relations; I deal both with the security dimension of a nation hosting a population which leads a marginalized life, and with the efforts of these people to secure a livelihood and a decent life.

Most of the political boundaries between the East African countries are drawn through semi-arid or even arid tracts of land. The boundaries have been disputed between the nations in times when more precise definitions of territories have been important. For the local populations this issue of nationality has been less significant. Cultural and ethnic groups are found on both sides of what are locally considered as artificial boundaries; to payment of goods can be made in the currency of one country or the other, etc. For some traders and livestock herders the boundary is even an asset in that it creates price differences and opens up opportunities for trade flows which would not otherwise occur. The freedom of movement across boundaries has to be there as long as extensive methods of livestock herding are practiced; seasonal migrations to customary pasture lands is one foundation for an entire production system. A collapse of this system would give rise to a major security problem with substantial numbers of impoverished rural dwellers moving to towns and cities. Political conflicts within closed boundaries have demonstrated this clearly.

In order to maintain a balanced situation it does not, however, suffice to keep boundaries open for the pastoral livestock herders. The herdsmen form part of a rural system. Pressure is strong on them, not least from farming communities with better watered land as their home areas. The forms of social structural adjustments to an inherent conflict over land use practice, relating to decisions whether to farm or herd, vary greatly. We often see mixed strategies. Many practices are situational and of a short duration while others are more lasting. Labeling first without a comprehension of empirical content does not bring a deeper understanding. It is even confusing at times to introduce labels such as agropastoralist for some of these practices. This is one of the points of my paper; that one must have a certain degree of first-hand knowledge in order to comprehend the processes in semi-arid eastern African lands.

I do not however, wish to plead for empiricism without theory. On the contrary, the first-hand knowledge must contain certain information; there are principal problems that reappear in different form. These structures, forming part of living conditions in the area, can best be gotten at through comprehending the perspective of the people concerned. With a community perspective the everyday problems can be listed. In a sense we have here the other dimension of "security"; the security needs of the individual indicate the problems he or she experiences. By comprehending these needs we form a holistic picture from a community perspective before moving on towards dealing with problems on a more general level. This is perhaps mainly a matter of method listening to the difficulties experienced and placing them in a wider framework. But it is a practical way of linking micro with macro perspectives.

Just to illustrate: a macro perspective may say that a certain piece of land is degraded through overstocking, that the area has to be de-stocked and tree planting introduced. Going ahead on these indications would most likely mean that the poor strata of a population would lose their remaining livestock and became dependent on wealthy house-holds, and that the victims of an unfair development process, not least the women, would not only suffer more but also be the ones doing the actual tree planting. A local perspective may add to the picture the political pressure from various social groups in an area, thus showing how land degradation relates to local economic, political and social structures. It is necessary to have this kind of competence, otherwise a local security problem may easily turn into a national or even inter-national security problem.

THE ECOLOGICAL LINK; FAMILY HERDS AS A RESOURCE AND AS A MEANS OF EXPLOITING RESOURCES

The major aspects of family herds as a link between ecology and society are that they not only exploit the vegetation cover in the area where man can hardly make a living on assets other than livestock, but also they comprise the decisive resource for subsistence and survival of man. One condition for long-term stability in herd sizes and compo-sition of households is determined by ecological and biological con-straints. We are then concerned with matters such as the relationship between pasture (access to water included) and herd size, with pace in reproduction and mortality. Another set of circumstances is shaped by man's competence to keep livestock. In the latter case the relationship animal/man requires harmony between family herd sizes and the composition of households. For the individual pastoralist the long-term aim has to be to strike a balance so that pasture maintains the capacity to feed livestock herds at the same time as livestock herds have the capacity to feed households.

A common expression for the relationship between pasture and herd is carrying capacity. It represents the limit of the number of animals that a particular area can carry. In the most simple case the system is self-regulatory. First the upper limit for the number of live-stock is transgressed. Then animals die in great numbers due to poor access to pastures. Pasture land recovers and the animal population can increase. To the extent that overgrazing or overbrowsing lead to an irreversible land deterioration, the carrying capacity of a tract is of course decreased permanently. The literature on livestock keeping in arid and semi-arid lands very often departs from varieties of this kind of self-regulatory ecological model. Man's management and technology do not reveal themselves in the approach. We might just as well reason about wild animals.

However, the importance of man for the balance between pasture and family herd is decisive. Many decisions of collective as well as individual character have a direct impact on the interplay between livestock and pasture. Ethnic groups or clans can cooperate or compete over the same piece of land; they might keep different species of live-stock which supplement each other for pasture utilization. Differences between groups in herding techniques mean in terms of carrying capacity that we have to introduce a spectrum of that capacity. Even individual visitors may have a direct impact. For example, a person might decide to slaughter individual animals, migrate, or leave live-stock herding altogether as a major source of income. Perhaps he/she decides to lend animals to relatives and go to the city to seek employment. Perhaps individual household members are forced to move to secure their income elsewhere in order to off load from their own household the burden of providing them with food. Thereby however the household loses access to the labour force of these particular individuals, a fact which might have very negative repercussions for the herding organization and thus the efficient utilization of pasture lands. One obvious implication is that the remaining members of the house-hold fail to cover all pasture land, so that some tracts become over-grazed while others are underutilized. This situation is often the root both of desertification and of the bush encroachment of the savannah.

Case 1: Hussein Mohamed; camel herder in northeastern Sudan

The following case of Hussein Mohamed illustrates the complexity of the situation of a pastoral nomad in achieving economic and social security. The picture is very different from human adaptation to ecological conditions.

In the area between the Red Sea and the Nile Valley in one of Africa's hottest and driest parts live the Amar'ar Beja, a Cushitic-speaking people. Many subsist on camel herding in combination with sheep and goats, some sorghum farming and wage labour in the town. The natural conditions for farming could hardly be more difficult. Accordingly the ecological stress is more or less built into the system for the inhabitants, and their case illustrates what can be done when room for manoeuvre decreases and when the risk for disaster increases. In essence the principal approach is risk-spreading through a high degree (as far as access to labour permits) of economic multiplicity.

The problem for the Amar'ar is to maintain a balance between household and family herd. It is clearly expressed through a dispersal of activities in livestock rearing and other economic fields with the aim of keeping the family together. The quality of herding suffers readily as a consequence of this need to spread activities, but also in consequence of faulty decisions in an ecologically difficult situation where the distribution of rain fall might be rather variable.

Hussein Mohamed lives in tracts where it is very difficult to subsist on farming. Living conditions are geared to uncertain rainfall. The area has one rainy season per year. Aulib, the inland plateau has a rainy season from July to September, though most years with a rather variable rainfall. The weather along the edge of the coastal strip of Gunob, and in the mountains parallel to the coast, is influenced by the monsoon wind. Thus both the Red Sea Hills and Gunob hereby have a rainy season in November and December. Only a very limited area between Aulib and Gunob along the western slopes of the Red Sea Hills may experience two rainy seasons per year. At least every other year the rains fail altogether. The uncertain rainfall regulates the volume of durra that can be grown and also influences the reproduction of Mohamed's camel herd. Births occur during the rainy season.

Hussein Mohamed's parents were once wealthy camel owners. In the wake of a very serious drought disaster in the beginning of the 1940s they lost most of their camels, and now after recent droughts the family lives on the verge of absolute poverty. Many relatives have been forced to seek a supplementary income from dock work in Port Sudan. Others combine animal herding with limited durra farming. Hussein Mohamed himself uses his family herd to produce basic food (milk and meat), and also as "a bank" where the surplus is invested. The animals are also made use of for transportation. Basic food for the household is camel milk and durra. The camel is clearly to be preferred in comparison with cattle or farming in that it provides milk and hence human food throughout the year. The quantity of milk which is produced by the camels varies with different breeds as well as different qualities of pasture and with the age of the animal. Milk production is clearly higher than that provided by cattle under similar circumstances. The family also makes use of meat, hides and fat. Male calves are mostly slaughtered along with aged females. The daily meat consumption, though, is not met with slaughter of camels but of small stock.

The household of Hussein Mohamed is dependent on durra to feed not only people but also at times livestock. This durra is produced through seasonal farming along river valleys (khors); but often substantial forced purchases are necessary to provide supplementary food. The transport camels are then normally used to carry home the goods. Apart from this service, the animals are also important for transporting houses with their straw mats and other utensils during nomadic moves.

Even if only a minority of the Amar'ar today actually keep any camels at all, the animal still plays a key role in the very formation of a societal ethos. The herding especially has a strong emotional connotation. The mish'arib, those men who carry out the distant herding, are often young, unmarried men who lead a life independent of their fathers (who however maintain some control over their sons long after they have actually married). Men are said to be strong, virile and attractive due to their diet, almost solely camel milk. It is the right, and duty, of the eldest son to become a herdsman. When no labour can be found within the family circle it is obligatory to hire a herdsman from outside. Then the normal salary is a young male camel per year (or the equivalent in sheep if that is agreed upon). As an alternative the whole household might follow the family camel herd if other activities such as small stock keeping or farming do not hamper such nomadic moves.

In general, camel herding is the responsibility of the men. Women have a traditional role to play maintaining sheep and goats by caring for them and checking their health in the morning hours before they are taken out to pasture. At times, when no boys are available, women also herd the flocks of small stock. But they are not supposed to milk either small stock or camels, and it is quite unusual for them to join in caring for camels. Women's domain is the home and the activities around it.

The nomadic home forms the very basis of livestock herding. A few milk camels are kept in the vicinity of the settlement for the daily milk needs of the household members. If pasture availability is scarce the camels may be kept far away, the distances defined by the need for water. Then a "satellite" to the actual home would be organized. Hussein Mohamed has arranged his household thus and accordingly has access to the id'amer, the permanent home where women, small children and old people live, along with a few temporary settlements arranged in accordance with climatic and pastoral conditions.

Particularly since the three-year long drought during the early 1940s many of the Amar'ar have been forced out of their traditional niche of camel rearing. This process was speeded up during the severe drought disaster in the 1980s. Camels are no longer the dominant economic base in the area. The relative importance of sheep and goats has increased and many households are forced to take up waged labour in Port Sudan, either as the sole source of income or as a supplementary activity where primarily young men of the household may be engaged.

The towns, Port Sudan and to some extent Atbara, are decisive for the continued existence of the Amar'ar as a localized group. Their economic importance for Hussein Mohamed's security depends primarily on job opportunities there. Members of his rather large household find permanent or temporary employment in town, be it in full time or part time. Some relatives have become townsmen and maintain social but not economic relations with their home area. Others have economic interests to look after in the family herd or in farming. For them it becomes important to live as close to the family estate as possible since they wish to influence the management of the assets. Those who earn enough in town might also benefit from maintaining contact with the surrounding rural areas which produce food but also offer investment possibilities in the form of livestock.

Urban incomes for his rural household are included in a total resource system for Hussein Mohamed. The most important source of income is dock work in Port Sudan. Almost all imports to the country enter through Port Sudan. Without the city the Amar'ar society would most likely have been dissolved and dispersed to different settlement schemes in the Sudan. Even before the drought around the mid 1980s, as many as 15% of all Amar'ar were directly dependent on regular salaries from their own or relatives' work in the port. The number of seasonal dock workers was considerably greater; registered workers were systematically replaced temporarily by relatives in need of occasional income.

Work in the port is hard, and most workers are men between the age of 20 and 30. They try to save money in order to invest in livestock (or start trading) to build a life career after the age of 30. Most of them fail to achieve this goal.

This is an illustration of the pattern of resources available for the Amar'ar. A person like Hussein Mohamed seeks different kinds of income utilizing the interphase between rural areas and town. He has to spread his social network both in town and countryside. Livestock keeping is also dependent on this fact, and not only the quality of pasture or livestock. One may ask which came first, a too poor food production from family herds or a voluntary striving for a more urban lifestyle. The fact remains of course that herders depend heavily on their herds for food acquirements. But this case, as does the second one from Kenya below, indicates that the town plays a very important part in the resource system of nomadic pastoralists.

Balance between family herd and size of household

Total dependency of a household on family herds would mean an unconditional quest for sustainable units, i.e. situations where both family herds produce enough food to cover the consumption needs of people, and access to a sufficient labour force for proper herding. The following graph illustrates the connections. It shows a case where food offtake from a family herd increases in a linear fashion with the number of animals, while the need for labour increases stepwise with the need to form new subunits for management reasons. As can be seen the net result is that sustainable combinations can be found at several intervals. Creating a surplus or establishing a margin against disaster for the individual means both moving towards the upper part of each interval in the graph, and stepping from one sustainable combination to another. This is how a nomadic pastoralist would build his security.

Figure 1. *Correlation between numbers of animals and people in consumption and production perspectives*

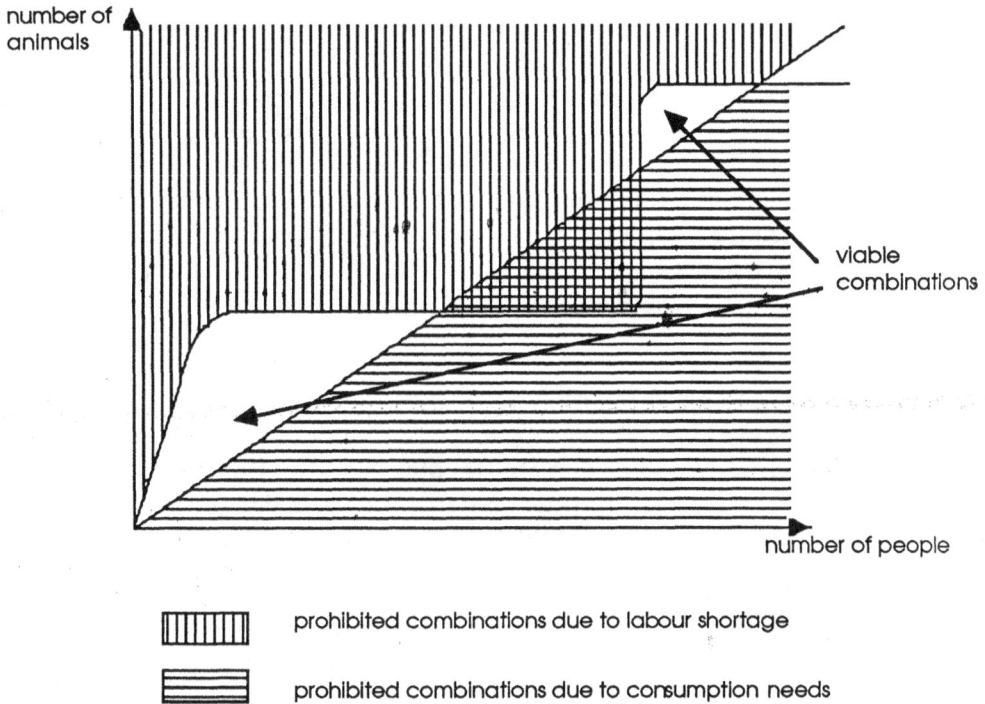

Source: Dahl & Hjort, 1979

Obviously, decision-making and management competence play a major role. The interplay between access to pasture and family herds becomes impossible to interpret unless one has more information on these matters. The picture which is often presented of nomadic pastoralism as a production system which in a self-regulatory fashion seeks a balance in (1) the relation between livestock and pasture, and (2) people to livestock, is misleading. What this kind of reasoning disregards is the fact that human effort improves the care of livestock. We may say that productivity is increased or that available natural resources are more efficiently used. With greater access to labour the possibilities improve, for example to separate herds into milk and dry herds, to reach underpastured areas, or to sub-divide livestock keeping so that one may manage herds designed either for meat or milk production, and keep different species.

75

Thus there is no unique natural limit for an area's carrying capacity. Instead one has to identify possible and non-possible combinations of the threefold connections man/family herd/resource exploitation. There are different levels of sustainable combinations under identical ecological circumstances. To put it another way: there is no one answer to the question of how many people a semi-arid or arid tract of land can support. What is stated theoretically is, for one thing, the absolute maximum solution and, for another, a range of possible and impossible combinations of the three "components" pasture, family herds and people. A competent herder manifests his competence through being more successful than his neighbours in allowing his family herd to grow. Thus he gains prestige since he thereby improves his capacity to support more people. The balance between the consumption and the production aspects of members of his household becomes less threatened. His security increases.

MANAGING RESOURCE SYSTEMS IN TIMES OF SCARCITY

The individual livestock herder devotes his attention primarily to the family herd. Pastures are often regarded as a collective asset which is not managed by the individual. The management influence tends to be indirect, operating over a long period through the influence of culturally prescribed rules on how to protect the young pastures (for instance those of the ethnic group or the clan) against overexploitation. In everyday life the offtake from family herds is entirely associated with the work input into arranging age and sex composition, calving rates and herding. Irregularities in age distribution, for instance shortage of heifers has to be compensated for through loan or purchase of suitable animals. Differences in reproduction rates provide an upper limit of possible biological growth of a family herd. The seasonal gearing of the time for calving leads to fluctuation in milk production, a fact which can partly be compensated for by maintaining different species of animals, given that one has access to enough labour for this. At certain times of the year work may be more intensive than at others. These and other complications have repercussions in determining how many individuals a family herd can actually support at a given point of time.

But also other strategies than proper herd management are involved in the resource management of dryland herders. The rural/urban interphase offers a temporary base for the impoverished pastoralists, rebuilding a family herd, or a permanent one for town-based pastoralism. Other security measures towards an improved sustainability are active social networks, managed in culturally typical ways.

Case 2: Letorongos and Lotukoi; poor and affluent town-based herdsmen in northern Kenya

This second case is linked with the former one in that it points to the role of the town in the resource system of nomadic pastoralists. Otherwise differences are great; a different kind of herding in a different historical context, a very small town and highly multi-ethnic circumstances. But the two examples have so much in common that many people try to maintain a presence in the livestock sector or try to return to it.

Different groups migrate to the small town of Isiolo in northern Kenya. They settle in and around town in order to gain temporary or more permanent subsistence. On the outskirts is Maili Saba which consists of a number of households which have all been pushed out of a pastoral economy in dramatic ways. Some have lost their livestock to cattle thieves, others have been struck by drought disasters. The inhabitants carry out farming on small irrigated fields along the Isiolo River. They own too few animals to support themselves solely on livestock rearing. But they strive to rebuild family herds, and in the meantime farm for subsistence and sell limited quantities of milk to the daily market in Isiolo town some ten kilometres away. Many of the households are headed by widows whose young sons have to make those decisions concerning husbandry that are normally made by elders. All young men who belong to the households of Maili Saba are more or less permanently elsewhere.

The houses in Maili Saba are built in traditional Samburu style even if they are slightly larger and more comfortable. Many pastoral Samburu who come here express their surprise that the village is permanent and that the inhabitants have made such an effort to build fine houses in spite of the fact that they are poor. A number of deserted houses indicate that some have left the village (no Samburu can inhabit a house that has been deserted by somebody else). None of the households in Maili Saba have any kinship relations. In Samburu talk the settlement is called Nairobi since it resembles the capital city with its mix of inhabitants from different areas.

The major part of Maili Saba is made up of of ten independent households. One of them belongs to Letorongos. Its inhabitants migrated here after great losses of cattle caused first by attacks by a former guerilla movement and then as the result of a severe drought. Left basically with number of goats Letorongos realized that he had to find other income sources. He decided to move with his family to the vicinity of Isiolo in order to try to rebuild the family herd using the rapidly reproducing goats. In the meantime he intended to support the family through farming. With his proximity to the town he counted also upon getting access to disaster relief. Earlier he had heard rumours about the availability of good pasture and proper water facilities in the vicinity of Isiolo. Raids by neighbours of the opinion that pastures belonged to them would create the only problem.

Letorongos' household consists, apart from himself of his wife Nolayiok and their two sons and two daughters. Both sons are employed as policemen in Wajir since they believe that there is no future for them in the livestock sector. The daughters live with their parents. The sons regularly send home a little money. Nolayiok and her daughters take care of the farm where they grow maize, beans and potatoes. In order to grind the maize she goes to the town. Some of the maizemeal is sold along with some of the milk from the animals. The income is just enough to purchase tea and sugar. Letorongos devotes his attention to the livestock. He has managed to build a basis for a new family herd, consisting of ten head of cattle in addition to a number of goats. Letorongos calculates that it will take a few more years before he can return to nomadic lifestyle.

The example of Letorongos illustrates the kind of options that remain for a herdsman in times of prolonged droughts when risk spreading within social networks no longer functions as a protective mechanism. This kind of situation has become more common when access to pastures becomes more and more limited in Kenya. In the neighbourhood of Isiolo, for example, animal parks have been established to protect wild animals, and land has been set aside for commercial activity. Furthermore, irrigated farming along the river beds has been established quite recently, and former open pastures have been fenced. All in all much of the dry season pastures of local populations have been lost.

Isiolo town however has not only become a possibility for survival for the pastoralists in the surrounding areas. It is also significant for providing a career which can never be achieved in the local community. Lotukoi offers an example of this fact. He is an old Turkana man from the upper part of Kerio Valley. At the age of 16 he walked to Isiolo together with a group of people, escaping the harshness of Turkana district and seeking jobs as road workers when the road between Isiolo and Garba Tula was built. Managing the small amount of family livestock had been a heavy task for Lotukoi, especially during rainy seasons bringing hailstorms which were both unpleasant and which dispersed the animals. He therefore immediately decided to leave the difficult life back home since he had heard that "Isiolo offered lots of food at cheap cost", a place where even the poorest of the poor could earn his living. Thus he fled from his childhood family which had so few animals that he was forced to go hungry most of the time eating "long'onyo", a kind of emergency food made out of dome palm nuts.

After the road work Lotukoi was employed by one of the town Somali as a herdsman for some cash income (10 shillings per month) and one cup of maizemeal a day. When he finally decided to return home he was able to purchase four bags of tobacco and a donkey. He traveled with this load through Baragoi and Maralal to Kerio Valley. Along the road he exchanged tobacco for goats so successfully that he decided to become a trader and sell tobacco rather than herding animals, an activity which he actually hated.

Today Lotukoi is well established with a livestock herd and several goat flocks back home. These are managed by relatives. He has, over the years, built a contact network which includes all the small towns between Isiolo and Kerio Valley. He has extended his trade activity to helping wage labourers in Isiolo to invest money in livestock in Kerio Valley. The way Lotukoi expresses this is that he borrows money from them. In return they get livestock from him which he allows to be herded along with his own animals until the time the workers return to Kerio Valley.

The case of Lotukoi is worth noting here also in that he is an agent between town and pastoral life. Resources within the monetary economy are converted into resources in a subsistence economy. In the rural areas there is, in a general sense, a shortage of money, i.e. livestock prices are too low. One result of people like Lotukoi, operating between the systems, is that of a redistribution of livestock ownership which does not harmonize with established patterns. A new possibility has been introduced whereby a man's social position can be improved through wage labour for the purpose of building livestock herds. This of course causes increased stratification, with a more successful minority owning a major part of the area livestock herd, employing herdsmen to care for this capital. This is an important tendency to note if a development strategy aims at countering such stratification; the causes of the phenomenon are not to be found locally but rather in the small towns.

Cultural variations in the formation of security networks

The balance between what family members consume and what they contribute to livestock rearing is a rather obvious problem in everyday life for most pastoralists. The external circumstances are difficult, at times with rain shortages and thus prolonged periods of drought. One consequence is that family herds at times exhibit high mortality rates with the result that the households' carrying capacity can no longer be maintained. Under such circumstances the spread of risks is attempted through different culturally varying principles for redistribution both of people and livestock. Members of households, which are no longer viable in terms of food production, can move to more successful relatives or friends, who have a need for more labour or whose herds at the time produce a surplus. Livestock are relocated to such households that have been hard hit to the extent that the collective group still has access to animals.

There are thus culturally different institutionalized ways of spreading risks, to form a security system. One such can be to build contacts between stock friends on an individual basis. Such a pattern is typical for the Turkana in northern Kenya. Here an adult male strives to establish a group of friends, geographically spread. Thus he achieves protection against a local natural disaster by establishing the possibility of borrowing individual animals elsewhere. Access to animals in such networks decreases to the extent that new and more individualized forms of ownership are introduced, such as the case of Lotukoi illustrates.

Another institutionalized way of decreasing risks is given by the principles for redistribution of livestock through social groups, primarily, or at clan or sub-clan level. This is a common practice among pastoralists. Here it becomes a collective undertaking for every member, in times of need, to support individual members who are hard hit. Membership thus means both the obligation to give support and the right to be supported in times of crisis. The effect is highly system protective; man emphasizes ethnic and even tribalistic memberships for the sake of upholding economic security through membership. In economic terms a redistribution of capital takes place, the off-take becomes low, and access to risk capital is nil. This indicates a major problem in development strategies, where risk distribution stands against economic growth capacity.

Marriage relations between different families provide further examples of social bonds which can be helpful for the redistribution of livestock. The bridewealth which the husbands' family pays to that of the wife varies very much in different cultures, as does the strength of later demands in cases of need for more animals. In certain cultures marriage means a one-time transaction in economic terms, while in other cases it means a life-long agreement that should the family of the wife experience economic difficulties it will receive further support. The Samburu, mentioned in case 2 above, illustrate the latter dimension. Here the level of bridewealth is low, around ten head of cattle, but there is also a built-in expectation of future help. The Turkana, also from Kenya, provide an opposite case; a high bridewealth of around 50 head of cattle, but then no expectations of further transactions.

The total effect of the "flow" of subsistence resources, i.e. domestic animals, in different social networks, becomes an obvious equalizer as long as one can disregard "external disturbances". During good years the pastoralist invests socially through his generosity in the security system (insurance) from which he will most likely benefit during a later period of his life, when he might be the one to be left with a decimated family herd.

This system means that "ownership patterns" of animals are multifacetted. Those animals which move in a family herd might belong to the herd of the household, to other household members, be borrowed, or be promised to somebody else. Milk production may be for the household as a group, or for one particular individual. The rights to food (milk and at times meat and blood) follow a social pattern. In this way every individual animal symbolizes a social relationship. If a genealogy of domestic animals were constructed, indicating where the offspring were located, one would get a map showing important social relationships expressed through gifts or loans of livestock (Dahl 1979: 92 ff).

Risk spreading for the pastoralist in arid and semi-arid areas can thus be seen as one general principle of social behaviour. The effect of an ongoing exchange of animals results in increased security in food production for the individual person. The ways in which the circulation takes place exhibit cultural and social differences. And the prize for risk spreading seems to be high with a narrow economic return.

SECURITY, SOCIAL MANAGEMENT OF PROPERTY AND ECOLOGY

Ecological circumstances may have an influence on the level of social organization (and hence by implication on the political structure), a fact which may be illustrated by a brief comparison between northern Kenya and northeastern Sudan. In northern Kenya large groups of people practice a nomadic lifestyle based on cattle rearing. In the case of northeastern Sudan, the Amar'ar have a distinct choice between a nomadic and a sedentary living. Even if the two territories are to some extent classified as being similar with regard to rainfall quality, there are clear ecological differences both in terms of vegetation cover and access to water. In the Amar'ar territory bushes and occasional trees are the main growth while an undervegetation is almost totally lacking, except during rainy periods. The territory is highly suitable for camel browsing but not recommended for cattle rearing. There is only one rainy season since the area is so far north. When the rain falls, the level of ground water rises in the normally dry river beds. This provides the conditions for a certain degree of farming over a short period (three months per year).

In spite of the fact that northern Kenya has a greater rainfall, with a more even distribution of the rain and with a less intensive heat, people here practice a more nomadic lifestyle than in northeastern Sudan. The fact that the rains fall irregularly and that cattle are herded, being watered every third day, brings a great need for mobility. A correlation between this need and collective ownership form (on clan level) can be found. The Amar'ar in contrast have a more marked individual ownership (small kinship groups), a fact which coincides with, it would seem, a highly marginal farming system where a successful harvest occurs perhaps one year in three. The harvest, on the other hand, is easily stored in the dry climate. In pits, which are dug in the hard laterite soil, the sorghum harvest can be stored for four years or even more.

The need for mobility in northern Kenya requires vast social networks both in order for information about the quality of pastures to flow, and to have access to relatives wherever one travels (the total stranger is not actually pushed away but is not dealt with in a friendly manner). The effects of marriage systems (which exhibit cultural variations between for instance, the Borana, the Samburu and the Somali) are also that kinsmen are spread over vast territories. The Amar'ar provide a sharp contrast in this respect. Here is a system with cousin marriage which means that the property is kept within the family group. At the same time we may note that wide alliances are not required from the point of view of herding practices.

In both localities there is a need to spread risks since harsh lands are inhabited. In Kenya it is often precisely kinship which becomes important for economic security. A person will strive to build vast kinship networks which have the effect of redistributing livestock towards those in the group who are worst off. For the Amar'ar this possibility is hardly available. Instead there is the advantage of an extensive, but over a long period of time reliable, farming system, apart from an old merchant tradition and a camel-based economy.

A herd of camels has a very slow growth rate but is resistant to drought. In a shorter time perspective it could be seen as an economic asset not too dissimilar from a farm. Establishment in a camel economy becomes very costly, and a herd represents considerable capital. To the owner, however, the offtake from this capital is what provides subsistence. An economy based on cattle is to a much greater extent a growth economy. This difference between cattle herders in northern Kenya and camel herders in northeastern Sudan also harmonizes with differences in kinship systems. It is important for the latter to keep together that property which is formed by a family herd of camels.

A certain difference in terms of ecological circumstances thus has an influence on what kind of livestock is kept, an issue which in turn also has organizational repercussions on a societal level. The animals have different pasture and water needs and require different inputs of labour for proper care. The low growth rate of camel herds implies a need to keep viable units together, a fact which can be traced in inheritance rules. Mobility varies. Camels are more mobile in their daily browsing but require fewer long trekkings for the Amar'ar. This means lesser needs for vast social networks and more of a need to defend effectively pastures at home.

These illustrations show how ecology, and to a certain extent biology, may have an influence on a societal level. There are also differences in technology and know-how, for instance the fact that the Sudanese pastoralists can grow sorghum in a way which is not within the competence of their Kenyan colleagues. The Amar'ar also have a vast amount of knowledge about camel rearing. We may distinguish a certain correlation between ecology, technology and societal forms but interpretations can never be unilinear. What can be achieved is a picture of unlikely and possible societal forms under given circumstances. The social management of property has to consider ecological circumstances. The individual security is established through this management.

Thus, we come back to the issue of room for manoeuvre as an expression of individual security. Marginalized groups (in a geographical, economic, political or social meaning) comprise individual households which have failed to build their security. The aggregate accounts for a major proportion of national and international security problems in connection with environmental degradation. It becomes necessary for individuals to pursue an imbalanced resource utilization. Such imbalanced production, often in combination with political stress or conflict, leads to inequality and political conflicts. Many groups and levels are involved; local and regional groupings in a country, nations and development aid organizations. We are concerned with a process where natural resources are utilized in a faulty manner so that in consequence a local population, hard up due to external circumstances, can see no alternatives. The basic question is to what extent man-made ecological degradation can be seen as a sign of social inequality and political stress.

CONTROLLING RESOURCES; A POLITICAL CONNECTION

Many stories about land grabbing in the better-watered areas and in high quality pasture-land are told. This is considered a "hot issue" but will not be dealt with here, except as an illustration of how a national economic and political system may reach "out" or "into" community based production systems. Society-at-large is present on the local level in a number of ways which often can be destructive in the local perspective. Cheap purchase of primarily farm land by city dwellers is but one of the more conspicuous examples. Other examples from the Kenyan "Case 2" vicinity are the encroachment on traditional pasture land by tourism (national parks and similar arrangements which hamper the movement of family herds), by veterinary administration (a large holding ground belonging to the Livestock Marketing Division) and by large development schemes preoccupied with irrigated farming. Water access, however, creates less of a problem.

The aggregated impact from national economic and political structures is considerable. The exploitative character, when seen from a local perspective, is significant but also so institutionalized that it becomes difficult to question. It can be illustrated with a series of maps on yearly migrations. In the Samburu case for northern Kenya the routes become smaller and the area covered diminishes every year.

This phenomenon, though significant, is slightly supplementary to the process I deal with here: The political development in northeastern Sudan and northern Kenya exhibits land degradation which can be linked both to such national processes but also to more local political developments. The Kenyan case illustrates the classical confrontation between farmers and herders. The common form it takes in northern Kenya is competition over pasture between nomadic pastoralists and herdsmen coming from farm lands further south. The latter keep cattle primarily as capital, while the former must maintain a higher quality in order to guarantee milk production. Thus an unequal competition emerges since the "agropastoralists" can exploit the pastures harder.

In political terms this competition follows ethnic lines. This is of course contrary to official policies which work against "tribalism". It comes out in local elections. In spite of the one-party system in Kenya local political rhetoric circulate around ethnic belonging and localized groupings. The nomadic pastoral groups in Kenya exhibit a weak political structure in comparison with neighbouring farming groups; these have more active networks to Nairobi and other major cities.

The other "classical" level of local political conflict is between different pastoral groups. Such conflicts are often centered around control over water holes. There is a range of culturally "prescribed" patterns for interaction or competition.

In northeastern Sudan the situation is quite different. The political structure of the nomadic pastoralists is more hierarchical and hence more of an asset in the nation-building political structures. Local groups' representation in Khartoum is strong enough to allow at least for efficient information flow. The environmental conditions are harsher than in Kenya, and traditional pasture control more complex, with a combination of customary rights and temporary rights. Water control is a key issue. There are no ethnic groupings as in Kenya, but instead a complex set-up of clans with limited land rights, with alliances and conflicts which evolve over historical time.

Both cases give ample evidence of the fact that control over natural resources is a major political issue, both for local politics and for the political effect on the local scene. The struggle for resource control extends also beyond natural resources and livestock. A major asset is information; for instance being informed in time about a coming development project could be significant. It is often utilized by local leaders, be they politicians or other opinion formers, for the advancement of their political careers. Another asset is of course control in a general sense over labour.

Acute local political conflicts occur in times of ecological stress. Then, as also during local elections, antagonism over resource control comes out in the open. Drought periods for instance, normally mean increased tension in local communities. If they are prolonged or natural resources significantly degraded, the issue at stake easily becomes whether or not to survive "back home". The political structures at national level, in contrast, are not so directly influenced by a prolonged drought. This is because the political process in the national context is more of a power game over national gains than over resource control in a particular locality. A wealthy trader's or politician's land purchase, for instance, may make an already fragile production system even more vulnerable. But it is more of a constant parameter to the system than what a temporary "land grabbing" by another outsiders' group would be. Thus we may hypothesize that the political connection with ecological stress is in local politics rather than the national scene. Only as a next step, if a generally chaotic situation emerges, so that there is a threat to national stability, are there considerable political consequences seen from a capital's horizon.

CONCLUSION: SECURITIES AS MEANS TO CONCEPTUALIZE SUSTAINABILITY

In focus for the discussion has been the security of the individuals living in arid and semi-arid tracts. I have tried to focus on times of ecological stress and political conflict. With the approach I highlight individual security, and not state security. As a footnote one must observe, of course, that the two do not coincide. Departing from a local perspective and seeing how problems appear for an individual is one way of formulating a holistic view; the one "offered" to a person actually "living in" a development process. This would be the micro perspective; the issue would be how it relates and interacts with a macro situation, on a state level.

Political international boundaries provide a drastic example of conflicting interests. In the local perspective it might be essential that they are kept open. In a national one it may be equally important to regulate movements across boundaries; they are not always relevant for migration patterns, and may even be beneficial for traders of various kind, and flows of assets. Closing of boundaries may lead to increased insecurity. Thus, political conflict leads to more insecure migrations and hence increased environmental degradation.

This kind of conclusion is of course self-evident. It illustrates the linkage from political conflict to ecological stress. But it does not automatically lead to another conclusion, that an overstocking problem should be attracted through a general destocking programme. What I have sought to illustrate by moving into details, is the complexity of production or survival systems. They must be understood, at least in outline, for proper planning. There seems to be no way around establishing a certain competence at the macro level about micro situations.

When we talk about environmental degradation we should not become too technical and leave out living conditions of human beings. We are after all, to a great extent concerned with a poverty kind of situation. This is to me one reason why an extended version of security can be so helpful.

In the drylands providing empirical evidence for this study, extensive livestock herding is a major undertaking. It is important for our own conceptualization of the situation to make note of the fact that those resources which people *manage* are family herds rather than pastures. This statement pulls the discussion to the old "tragedy-of-the-commons" debate; i.e. "private" herds kept on collective land. I do not enter into it in the paper, though. In essence I consider the notion too shallow, and rather like those of carrying capacity, stock units, over-stocking or the like. They require more precise specifications, culturally or else.

The tendency to rely on macro statistics alone is dangerous; we are often presented information which becomes so to say a "piece of truth" tending to replace the more complex systemic understanding, which then is more tiresome to comprehend. Through insisting on a security notion we may keep a forceful demand not to forget the issue; that we are concerned with fellow man. In the destocking case; *where* should people go? And if we search for "indicators" of environmental stress, there may be more reliable ones at hand; a change of strategy could be shown in the setup of domestic livestock species in a household. Furthermore, a post-drought situation shows in herd age compositions, etc. When comparing camel herds in northern Kenya and southern Somalia, a few years ago, those in Somalia were composed of younger animals and produced more milk (Hjort af Ornäs and M A Hussein, 1988). This seemingly favourable picture is shaped by a previous drought disaster, against which the Somali herders were less prepared than the Kenyan ones in this particular case.

And speaking about camels let me provide one complicating example from the field of camel herding. I am thinking of a case where cattle herders decide to keep camels as well; the benefits have been (a) increased sustainability; (b) improved land management (clearing and better grosses); (c) increased food production (milk). Thus improved security and less land. But statistics say: higher stocking rates.

I have in this essay discussed the interrelationship between family and household sizes on the one hand, and herd structures and sizes on the other. Security is built through proper herd management. This means of course large herd sizes, but at least as important are proper age structures, and proper mixes of different species. If we should begin expressing views on over-stocking, this has to be one point of departure; not merely a decrease in numbers, but structural changes in family herds. The issue is certainly located at a micro level, has both national and regional implications, and concerns security. One major message is of course that unless this kind of individual, or local level, security is accommodated, other levels of security are threatened even to the point of military security. I have in this essay provided examples of culturally different ways to seek security though formation of social networks.

In the essay I have also tried to look into the non-sustainable situation. Both cases quoted bring attention to the role of an urban development in rural context in Africa. I want to propose that this is not specific to these micro cases, but rather a major phenomenon not least for nomadic pastoral populations. The phenomenon of small towns located in a rural context is a major factor both expressing a lack of security, an imbalanced environmental and political situation, and at the same time offering a possible means towards improvements. In this respect one could claim that small towns could be added to the list of "indicators" of stress. And they not only indicate such stress, but also adds to the degradation, unless proper measures are made.

Finally and by conclusion, I have ventured above to have a look at political conflicts, fueled by changes in resource control with, as I suggest, a kind of flip-flop quality: For instance, less access to seasonal pastures leads to increased ware on land, decreased security, and growing local conflicts. In times of ecological stress inherent local conflicts, often with tribal connotation, not only comes fore more distinctly, but they also threaten to turn into major national and international issues. Thus, also in a classical sense of security, the issue of sustainable development on a local level has a very direct bearing also for a possible armed conflict. Maybe we could hypothesize that future wars will be fought more over such natural resources which one as an outsider might not regard as valuable, such as extensive pasture land, which however prove to be absolutely vital for a local population.

REFERENCES

Dahl, G. 1979: *Suffering Grass: Subsistence and Society of the Isiolo Borana*. Stockholm: Stockholm Studies in Social Anthropology, University of Stockholm.

Hjort af Ornäs, A. and M.A. Hussein 1988: "Camel herd dynamics in southern Somalia: Long term development and milk production implications". In: A. Hjort af Ornäs (ed): *Camels in development*. Uppsala: Scandinavian Institute of African Studies.

Ecological Degradation in the Sahel: The Political Dimension

Abdel Ghaffar M. Ahmed

INTRODUCTION

The United Nations' dream that the 1960s would be the "development decade" never came true. Instead, many of the plans that were, then, nicely constructed for the Third World countries have had a dismal record of implementation. The preamble for the Lagos Plan of Action affirms that:

> "The effect of unfulfilled promises of global development strategies has been more sharply felt in Africa than in other continents of the world. Indeed rather than result in an improvement in the economic situation of the continent, successive strategies have made the continent stagnate and become more susceptible than other regions to the economic and social crisis suffered by the industrialised countries. Thus Africa is unable to point to any significant growth rate or satisfactory index of general well-being in the last twenty years" (O.A.U, 1980:3).

"Why did things go wrong?" is a question normally asked by bemused local and foreign experts, bewildered planners and a despairing local population. Had these plans achieved part of their specified objectives the continent would have been a different place today.

The human and natural resources of Africa are immense and are there to be tapped in the appropriate manner.[1] When this happens, if ever, the continent can satisfy the needs of its population and become a major contributor to the World economy. At present, and despite all of its potential riches, Africa has 20 of the 31 least developed countries of the world, and most of its inhabitants are under economic, social, political and ecological stress.

It has to be noted that most of the resources of Africa are non-agricultural and are unable to provide short-term sources for foreign exchange that can help Africans to bridge the gap between their capacity for food production and the number of population to be sustained. The issues of dealing with environment, and the proper management of existing resources, as well as the appropriate matching of imported technology with existing indigenious technical knowledge are the ones that need to be comprehensively addressed by experts, planners and local inhabitants alike if any real development and easing of the present stress in rural areas is to take place. The cultural dimension in development planning in African countries has to be reconsidered. Emphasis has to be more given to those elements of culture that relate to values, systems of beliefs and various aspects of social organisation. That is because it is in these elements that possible solutions for any conflicts arising from stress may find their roots.

Africa, and especially its Sahelian zone, is now facing one of its most serious disasters ever, namely, the famine. This disaster has been at some point attributed to meteorological catastrophe and a general ecological degradation of the continent and specially its Sahelian zone. The reason for the Sahelian challenge is summarized in the lack of capabilities of its countries to deal with deforestation, desertification, drought, population growth and hence the limited food potential resulting from these factors combined. The situation in the African wet lands and flood plains results from the presence of excessive water that covers wider areas of land and make it out of access for most of the year, thus being inadequately utilized by human and animal populations and hence its food potential is also limited.

Man's intervention in the form of over-intensive farming and/or overstocking of animals, in more accommodating areas of the continent, without close consideration of the carrying capacity of land was, and still remains, a major contributor to the continent's disaster. Ecological stress became one of the main features of most parts of the continent. In its wake an unwitnessed population mobility, much greater in magnitude than any seasonal migration that customarily takes place on the plain of Africa, has started. In fact such a mobility included both human population, domestic animals and wildlife who decided to leave their traditional areas that came under stress to travel to more prosperous ones. The result has been more ecological degradation in new areas. The civil wars which are currently in progress have added to the environmental factors to help aggravating the famine situation.

Policy decisions to remedy the situation and, in the process, to suggest appropriate plans for development have been a major concern for planners, experts and even, sometimes, local population. However, no distinguished cases of success can be cited. The main causes of failure have been so far attributed to the inability of most plans to take into consideration indigenous environmental and technical knowledge. Paradoxically, many of the most damaging decisions that led to rapid degradation and resulted in major conflicts have been introduced and supported by development experts, whether national or expatriates. In suggesting development programmes these experts seem to have given little thought to questions such as "who benefit from such a development project? or what impact would such a project have on local culture?" Nor did these experts bother to list some of the reactions of the subjects of their well constructed programmes. The channels of communication were all the time uni-directional and are from the top to the bottom. Participation of local people as equals in the development process was rarely considered and no use is made of their indigenous environmental and technical knowledge.

LOCAL AND REGIONAL ELEMENTS IN ECOLOGICAL DEGRADATION

Until a few decades ago the Sahelian population used to produce surplus food commodities. A few years ago, with the drive towards cash crop production generated by relations and mechanisms that linked this region to the world economy, this capacity for producing a food surplus declined. Today the surplus of past decades has turned into deficit which, according to the projections of the FAO, is to increase seriously by the year 2000 (FAO 1980).

Even though it can be accepted that climatic factors have, to a large extent, been responsible for the increase in the magnitude of disaster, man in general has a major contribution to what is presently taking place. Indicators which show that such a disaster as the one in the Sahelian zone can happen were identified long ago in the case of some countries. When looking at the case of the Sudan it can be seen that as far back as 1943/1944 some warnings were aired concerning the development in the central part of the country. It was pointed out that the trends of expansion in cultivation of crops in rainfed areas, the cutting of wood for energy provision and opening of boreholes in different areas have to be monitored carefully or else the degradation of the environment will be difficult to control (Sudan Government, 1944). The sand dunes movement and the desert expansion in a southern direction was a worry to those concerned. This being the case, however, no serious measures were taken at the central government or the village levels. Nuclear and extended families in marginal land, whether in settled villages or maintaining seasonal migration with their herds of animals, dealt with the natural resources of their areas as if they cannot be exhausted. The literature on herds increases and the increase in marginal lands which were put under agricultural production is enormous.[2] Yet it is important to emphasize that in opting to take such a line of action, local inhabitants in rural areas, whether at village or camp level, had to make ends meet in the new economic setting that emerged after independence (i.e. after 1956). Nor were successive governments, for that matter, free in their decisions. They were bound by the local realities of life and hence, most of the time, were forced to bend for local political pressure to dig new wells in different areas so as to avoid conflict between people in different localities and at the same time to gain political support in these areas. The local traditional leadership worked for the fulfilment of the demands of local people in order to maintain its power and authority positions.

Neither the government at the national or regional level nor the local inhabitant questioned the wisdom of their actions in the utilisation of the available resources and did not give real consideration to the possible consequences. The degradation and stress on the carrying capacity in grazing areas and the destruction of the top soil came gradually, almost unnoticed in the savannah belt of the Sudan which represents the eastern extension of the Sahelian zone. It is true that in the wake of the Sudan-Sahelian drought of 1968–73 the Sudan was often cited as the only exception to the then widespread mass starvation. This was attributed to a pattern of agricultural development during the late 1960s, in rich clay soil leading to the expansion of capital-intensive food production aimed at supplying internal markets rather than the export markets. After the implementation of the five years plan of 1970–1975 the planned area under mechanised rainfed agriculture increased by 1,308,000 feddan which represented a total increase of 66% in terms of area, not counting the unplanned extension which is by far more than this percentage.[3] This increase was reflected in a 72% rise in the production of sorghum, the staple cereal for the majority of the country's population. One consequence of this unparalleled story of success at a time when the whole continent is struggling to help its famine victims, made planners and investors start to consider the country as a potential area for further agricultural expansion. However the suggestion and encouragement were directed to cash crops for the export markets. Due to the rapid expansion that took place in a short period, and the policy of land mining which was adopted by owners of large schemes in rainfed areas, the soil erosion started to take place and the crop production started to decline (Abdel Ghaffar, 1976).

The expansion in rainfed mechanised schemes also took over land which was traditionally utilized by nomads (Abdel Ghaffar, 1976, 1978, 1980) and hence decreased the area available for grazing as well as driving villagers out of their traditional villages and turning them, in some cases, into wage labourers. This competition for land led to conflicts between nomads and mechanised rainfed scheme owners. Nomads had to fight their way through these schemes in their seasonal migration since the schemes have taken areas where the nomads traditional migration routes used to cover. This also led to competition between settled villagers and nomads who were in the past maintaining a relation. The situation was even more complicated when, by the end of the 1970s, nomadic and settled villagers from marginal areas of the semi-desert region started to move south with the hope of avoiding starvation for both human and animals.

Politically the result was lack of stability in the whole region, reaching in many cases the point of tribal wars due to conflict of different management systems used in relation to the available resources. This can be seen in the continuous conflicts between the Bani Helba cattle herders and settled villagers and the Rizegat "Jammala" (camel herders) of northern Dar Fur who had to move south to avoid the consequences of the drought. Since 1976 these two groups have engaged in conflicts which have claimed many lives.

Another dimension of the ecological degradation has been the mobility of various groups in what seems to be a continuous migration between rural areas. Sometimes this extends beyond the international boundaries of some of the Sahelian countries and adds the category of drought and famine refugees to the already existing large number of refugees from the civil wars that became a major feature of the political life in these countries. There is also an increasing evidence of migration to take refuge in urban areas in order to avoid starvation, as can be seen in the light of what is happening in the Sudan now. A large number of rural areas have been abandoned and the local population have moved to the towns on or near the banks of the Nile, or, in many cases, towards the capital, i.e. greater Khartoum. This movement has evidently created an extra demand on the state structure which is not equipped to accommodate such numbers in the urban setting. The state apparatus has failed to handle the situation and political chaos and economic disaster seems to be the main feature at present in everyday life. Underlying this whole issue is the conflict of the perception of development in relation to the different strata within the society. Combined with the ecological degradation and the local political decision is the integration of the rural areas in a wider system that promoted the emphasis on "development from above" and hence the promotion of the elite's idea of development.

NATIONAL AND EXTERNAL FACTORS IN ECOLOGICAL DEGRADATION

The national economic policies of the Sahelian countries seem to be a major cause for ecological degradation and the political chaos resulting from the environmental stress. It is not the intention here to deal with the details of these policies. What needs to be noted is that when it comes to their national economic policies these governments were, and still are, not free to choose their lines of action. Their incorporation in the international economic system has ascribed them to the role of the raw material producers. They are caught in the dilemma of having to direct their rural people to produce cash crops so that in the end the country in question may obtain foreign exchange which it uses to pay for manufactured goods. These manufactured goods are forced on them by the dominant capitalist system. A certain mechanism has dominated the relation of these countries with the developed world: "The poor are constantly buying dear from the rich and selling cheap to the rich. There has been a regular transfer of wealth from the poor countries to the rich countries which appears in no statistic about aid, or even in those related to what are called 'net transfers' to Africa or other Third World countries" (Nyerere, 1986:389). There is no way to argue against the fact that African countries in general, and Sahelian countries in particular, have made mistakes in constructing and implementing their development plans. One of the obvious mistakes is a dependence on a top to bottom approach or insistence on this development from above without due consideration for local cultural dimensions. Yet, at the same time developed countries, as one African leader put it, need to be reminded that this is one world. "African poverty and underdevelopment is not unrelated to wealth and technical advances elsewhere. The existing pattern of wealth distribution in the world is the inherited (cross) of independent Africa, not the result of Africa's own actions. It is not irrelevant to the present condition of the Sahelian zone that one-quarter of the world population has four-fifths of the world's income. Wealth breeds wealth and poverty breeds poverty—through the relative investment capacity and power of the powerless in relation with others" (Nyerere, 1986:388).

The consequence of the slow development (or underdevelopment) of the Sahelian countries has been more and more involvement of international institutions in the policy formulation of these countries. Massive foreign investment in some cases can be cited (i.e. Kinana Sugar Scheme in the Sudan). On balance such investment has been geared towards the exploitation of existing resources rather than generation of development-oriented activities. Most of this investment does not seem to reach the suffering people nor does it attempt to involve them in the process of production. The small farmer has never been a target group except in the nicely worded documents that fill the shelves of experts, politicians and academics alike. The lands that the rural poor used to, traditionally, own were in most cases, as stated above, taken over and redistributed by government authorities to new investors, be they national or foreign.

It cannot be ignored that assistance and aid are being offered to developing countries. However, it has to be noted that though some of the intentions may be sound the reality of the situation says otherwise. A look at the World Bank support for the small farmers in the Sudan and its collaboration with the Sudan government in Kordofan gives a typical example of how assistance and aid can be made to serve an exploitation purpose. The farmer is financed, but only when he or she accepts to cultivate oil seeds. The World Bank jointly with the government-owned Agricultural Bank undertakes the responsibility for the harvest, i.e. receives it from the farmer and decides when and how to sell it and then settles the standing account. This has been an unhappy relation where farmers in most cases are indebted and have to work harder on the same terms, and borrow more from the bank and be tied further to these institutions (Fadlalla, O. et al., 1982). This can be seen as a micro-image of what happens to the state (Sudan) with reference to its links to the IBRD and the IMF (Brown, 1985, Ali 1985). Relief aid has been offered in recent years, and especially in late 1988, from various parts of the world to the African people under stress. The effectiveness of such aid in saving the lives of thousands of people in the short run cannot be denied. However, it is justified to ask about the long-term impact of such aid and in what way will it influence and shape the future of the local people who are receiving it? Most of local people who started receiving this aid seem to ignore the possibility of going back to their localities and start producing their own food even when the circumstances allow that. The aid agencies seem to be willing to provide more and more non-productive support to local people to an extent that rightly leads to the impression that these aid commodities are a surplus generated in developed countries and need to be dumped elsewhere. The relief is creating the necessary market for goods which are otherwise being rejected. It is another way of subsidizing some unsuccessful industries in developed countries by their own governments without risking public opposition.

In their attempt to channel their development assistance in the past and their disaster relief aid at present, many international institutions, with or without the support of national governments, have attempted to gather information through sending "experts" to conduct research in the target areas. The research activities in the Sahelian zone have been positively correlated with its underdevelopment. It can in many ways be said that the area has been over-researched. Could words be of use to hungry people, the population of the Sahelian zone would have been the most nourished people on the African continent.

Yet the present disaster has shown that some of the research results and the recommendations that followed from them hardly match the value of the stationary on which they are written. What is needed is not more research similar to what has been taking place over the past decades. The need now in the Sahelian zone is for action-oriented research that involves more than the mere professionals. Enough, probably, has been heard from professionals whose attention has been paid more to the results of their professionalism than to the applicability of the results of their research to problems facing the indigenous population.

It is not surprising that the "experts", who are normally urban-oriented elite, lack the sense of sympathizing with the rural poor. Their sympathy matches with that of foreign "experts" who offer to help their central government. Behind this lies the fact that these elites are the product of systems of education molded by the ex-colonial powers, the result of which are persons who by all standards are acting as agents of neo-colonialism.[4] Whether they are doing this consciously or not is a point beyond the scope of this paper. However, when these categories of "experts" sit together they can easily decide on the destiny of the poor population of rural areas. Aid donors, represented by their experts, and central government ministers and their advisers, when reaching an understanding on a project, be it a hydroelectric dam that displaces thousands of people or a road that cuts through forest lands, face few, if any, legal or political constraints which can hinder their decision (Report for independent commission, 1985:33). Experimenting on rural peoples' style of life became a hobby for the government advisors whether local or foreign. Africa has too frequently been a laboratory for the fancies of foreign development experts who always leave when a project fails in one area or one continent to try again elsewhere. This failure is counted as a relevant experience by international agencies. However, to blame what is happening in the Sahelian zone on the central government representative and the aid donors and their experts that came as part of the aid package, no matter what their quality is, would be a simplistic view. The local people as well as the global economic system are equal partners in what happens and should be equally exposed if something is to be learned and the disaster is to be overcome. The reality of the situation in the international scene is that we have two categories which can be characterised as "aid pushers" and "aid addicts" who equally need treatment.

CONCLUDING REMARKS

The ecological degradation, whether due to natural factors or as a result of man-made decisions at the local, national or international level, has affected the life of human and animal populations in the African continent. It generated unprecedented mobility and led to conflicts in almost all parts of the savannah belt. Management systems of natural resources that used to support each other through long lasting symbiotic relations have had to come in conflict and lead to the destruction of each other.

This issue of mobility of human population in the Sahelian zone, and the fact that small farmers have to come into contact with representatives of state-owned or internally financed institutions, has led to the promotion of the idea of individualism in communities whose work ethnics has mostly if not totally depended on collectivism. Even most of the resources are communally owned, and traditionally no person would attempt to take more than his or her share. There were few or no attempts to expand beyond the limits of what covers the necessary needs since production was mostly for consumption. Even consumption was in some cases a collective activity. The action of adopting the individualistic ethnics has since resulted in a total degeneration of community cooperation. Cooperation used to be one of the main features of the traditional production systems. In the recent past these societies used to have their well developed systems of collective labour and the indigenous technical knowledge to go with it. They no longer maintain that cooperation, and no collective responsibility towards each other or towards the natural resources seems to exist.

All the development plans in the savannah belt seem to have a total disregard for the nomadic population of this belt. The fact that they had traditional rights over vast areas which were taken over by private capital owners or given to state-owned corporations and internationally financed institutions does not seem to count. The government policies did not give consideration to the local population or allowed for popular participation of a significant magnitude in decision-making processes. The result was a state of frustration which led to local and regional conflicts which in most cases resulted in political instability and total chaos in most of the Sahelian states. Tension and aggression is a common feature of daily life in the urban setting of the countries of the Sahelian zone.

NOTES

1. "In addition to the reservoir of human resources, (Africa) has 97% of world reserves of chrome, 85% of world reserves of Platinum, 64% of world reserves of manganese, 25% of world reserves of copper, without mentioning bauxite, nickle and lead, 20% or world hydro-electrical potential, 20% traded oil in the world (if the US and USSR are excluded), 70% of world cocoa production 1/3 of world coffee production, 50% of palm produce, to mention just a few". OAU, 1980:60).
2. Most of this literature can be found in international and national journals specializing on Africa in general or development in Africa in particular. The "Sahel" journal published by Michigan University gives regular summaries on research and publication on the Sahelian zone.
3. The following table obtained from the Mechanized Farming Corporation in the Sudan shows the magnitude of expansion:

Locality	Demarcation	Area before the 5-year plan (1970–75) feddan	Area after the 5-year plan (1970–75)
Gedarif	1944	1,372,000	2,246,000
Dali	1957	156,000	104,000
Masmoun	1957	198,000	188,000
Renk	1964	120,000	343,000
Habila	1969	123,000	396,000

4. For further details on the role of the elites in one of the Sahelian zone countries, see Abdel Ghaffar (1979). What is described for the Sudan in this case seems to be a typical example of the elite, no matter which of the European colonial powers was in control.

REFERENCES

Abdel Ghaffar M. Ahmed, (ed.) 1976: *Some aspects of Pastoral Nomadism in the Sudan*, Khartoum.

Abdel Ghaffar M. Ahmed, 1978: "The Relevance of Indigenous Systems of Organisation Production to Rural Development: A Case from Sudan", *Essays on the Economy and Society of the Sudan,* A. Hassan (ed), ESRC, Khartoum.

Abdel Ghaffar M. Ahmed,1980: "Development Planning and the Neglect of Nomads: The case of the Sudan", in Haaland, G. (ed), *Problems in Savanna Development,.* University of Bergen.

Abdel Ghaffar M. Ahmed, 1979: "Tribal Elite: A base for social stratification in the Sudan", *Toward a Marxist Anthropology.* S. Diamond (ed.), Mouton.

Ali A. Ali, 1985: *Sudan Economy in Disarray.* Itchaca Press, London.

Brown, Richard, 1985: A background Note on the Final Round of Economic Austerity Measures Imposed by Nimeiry Regime: June 1984 to March 1985, *DSRC/ISS Work Paper.* Univ. of Khartoum.

Chambers, Robert , 1983: *Rural Development: Putting the last first.* Longman, London, Lagos, New York.

Deng, Francis Mading "Development in context", in M.W.

Daly (ed), 1983: *Modernization in the Sudan.* Lilian Barber Press, Inc.

Fadlalla, B.O., Abdel Ghaffar M. Ahmed and M.O. El Sammani, 1982: *A Rural Credit Institution for Kordofan Region.* Regional Ministry of Economic and Finance, Kordofan, Sudan, Oct.

Independent Commission on International Humanitarian Issues, 1985: *Famine: A man-made Disaster?* Pan Books Ltd.

Nyerere, J.K. 1986: "An Address", *Development and Chance.* Vol. 17, Number 2.

O.A.U, 1980: *Lagos Plan of Action for the Implementation of the Monrovia Strategy for the Economic Development of Africa.* ECM/ECO/9XIV) Rev. 2, A/S-11/14.

Sudan Government, *Soil Conservation committee's Report 1944.* MacCorquodale and Co. (Sudan) Ltd.

The World Bank, 1984: *World Development Report 1984.* Oxford University Press.

Ecological Stress, Political Coercion and the Limits of State Intervention; Sudan

M.A. Mohamed Salih

INTRODUCTION

The recent processes of ecological imbalance in the arid lands have been the subject of many studies. The literature has so far highlighted the relationship between human activities, settlement and the utilization of natural resources. There are also some well documented studies on the relationship between ecological degradation and its disastrous impacts such as famine, floods, population flight and the like. [1]

However, in the particular case of the Sudan, studies pertaining to the understanding of the relationship between ecological stress and political conflict are scarce. A general critique has been directed against the subsequent plans for socio-economic development rather than the deficiency of the state in dealing with natural calamities and their consequences.

The state is generally described as an institutional mechanism for the articulation of the general public interest or as an entity representing the interest of a dominant interest group through those who yielded a grip on power. However, the main argument in this undertaking is that while the Sudanese state has been fully exercising and at

[1] For example, Bryson, R. 1973, *The Sahelian Effect*, University of Wisconsin Press. Derrick, J. 1977: The Great West African Drought, *African Affairs*, 76-Oct. and 1984, West Africans Worst Year of Famine, *African Affairs*, 83. Sen, A. 1981: *Poverty and Famine*, Oxford University Press. Spooner P. and Mann, H.S.(eds), 1982: *Desertification and Development in Dry Ecology in Social Perspective*, Academic Press. Watts, M. 1983: *Silent Violence, Food, Famine and Peasantry in Northern Nigeria*, Berkely, California University Press. Franke, R.W. and Chasin, B. 1980: *Seeds of Famine: Ecological Destruction and the Development Dilemma in the West African Sahel*, Montlair. Timberlake, L. 1984: *Natural Disasters: Acts of God or Acts of Man*, London, Earthscan. Redcliff, M. 1984: *Development and the Environment Crisis: Red or Green Alternatives*, London. The International Commission for Humanitarian Issues, 1985: *Famine a Man-Made Disaster*, London. Lawrence, P.(ed) 1986: World Recession and Food Crisis in Africa, Special Issue, *Review of African Political Economy*. Micael, H. G.(ed) 1987: *Drought and Hunger in Africa*, Cambridge University Press.

times exceeding its jurisdiction powers, it has so far failed to meet its duties and obligations towards its displaced citizens. In its present situation of underdevelopment, ecological stress has contributed to the transformation of the state objectives from development to bare management of survival. Furthermore, the holders of power perceived the survival of the state as being more important than that of its citizens. Since the state is incapacitated by the present economic crisis, internal and external pressures and ecological stress, the power holders were not able to justify their holding of power in terms other than guarding their interest or that of the dominant classes in the society. The whole objectives of the state have yet again been transformed from development and rehabilitation to mere maintenance of compliance and order. In other words the state policies have revealed an increasing tendency towards the use of repressive measures. This has, on the other hand, contributed to the persistence of certain totalitarian features in the handling of the emergency situation. Hence political coercion is used to protect the survival of the state vis-a-vis the interest of the general public and the survival of the victims of drought and famine.

The Sudan adopted since the 1940s the policy of using coercive measures to evict traditional cultivators and pastoralists from their farms, animal routes, grazing lands and water points in favour of the expansion of large-scale mechanized farms. When millions of the victims of 1983/1985 drought and famine crisis began to search for possible survival alternatives, they were constantly challenged by the state authorities which opposed their spontaneous movement into new fertile lands. The state authorities did not appreciate the movement of the victims of drought and famine into the wetter zone or some of the areas which have been demarcated for the future expansion of large-scale mechanized farms. Such areas were carefully monitored by the regional authorities against spontaneous settlement. The army was deployed in some areas to protect the large-scale mechanized farms and to monitor the movement of the small cultivators and pastoralists instead of encouraging them to settle in the wetter zone. The potential survival and coping mechanisms which could have been adopted by the victims of famine were, in this case, being intercepted by the state which made it difficult for them to continue their pastoral and farming activities. The immediate result of this was that millions of people began to move close to towns and relief centres only to live on international charity.

Large segments of the displaced and victims of the 1983–1985 famine began to move to towns and urban centres to be within reach of relief supplies. In addition to starvation, the urban refugees have confronted a new set of problems related to a) lack of employment due to the widespread economic recession and even when jobs were available, they did not have the skills or the education to take them. b) social problems related to the feeling of despair and uncertainty emanating from the psychology of hunger and deprivation. c) housing problems which culminated in series of confrontations with the municipal authorities, especially in the capital towns of Khartoum, Khartoum North and Omdurman.

The coercive measures introduced by the state have been stretched to reach the displaced and victims of famine in towns where the Town Planning Authorities considered their settlements illegal and used the fist of the police and the army to demolish their houses and repatriate them to the very conditions which mitigated their immigration in the first place. The political conflict here is one between the urban refugees whose main concern is with survival and the survival of those who hold power in the state amidst the fear that poverty and deprivation are potential sources of political uprisings and social revolutions.

The theme of this paper, therefore, is that the use of various forms of political coercion challenges the relevance of the legitimating ideology which perceives the state as a source of security. On the contrary, the state policies are conceived as an instigator of individual and group insecurity in the face of an increasing ecological imbalance and economic recession. Political coercion therefore has defined the limits of the state intervention in terms of its inability first, to address the immediate problems of underdevelopment and second, its failure to assist those who are incapacitated by drought and famine. Political coercion became the only available means for the state to impose its legitimacy and hence assure its relevance and existence.

THE STATE POLICIES AND ECOLOGICAL STRESS

The state intervention in the affairs of its population is as old as the concept of the state itself. Nevertheless, in this section, I intend to provide a brief account of the intervention of the state in ecology and society in the Sudan. It covers the period from the advent of the Turco-Egyptian colonial rule (1821–1885). This period is considered by some historians, for one reason or another, as the marker of the modern history of the Sudan. This could be true only if the concept of modernity is cynically linked with colonialism. However, what is true is that, the colonial state interfered in the affairs of the local population through the introduction of new administrative structures and the imposition of modern economic measures such as taxes and export crops; a process which started well before the 19th century.[1]

The attempts of the Turko-Egyptian rule to introduce cash crops, especially cotton, began in Kassala in 1841 when a 30 km long canal was constructed for irrigation purposes. An expansion in the Gash Delta followed in 1860 also to grow cotton and other food crops. These activities contributed to the concentration of the population of Kassala province in large settlements who were forced to grow more crops in order to be able tp pay taxes which were considered a constant financial obligations towards the state. High taxes were levied and paid in crops and slaves as money was so scarce. The taxes were so high that some villagers had to pay almost half of their produce every year to the state. In the bad years, heavy taxes forced many villagers to flee their farms to remote areas outside the reach of the Government. For those who decided to stay behind the tax burden demanded hard work and a steady expansion of the land under cropping. As the population was still very small relative to the size of the country, there were no reports of over-cultivation or over-grazing. However, the existence of a strong centralized colonial state and its ability to pacify various groups and bring them under its control meant that the state was already taking decisions which had far reaching impact on how the population should handle their economic life.

The Turko-Egyptian rule was defeated by the Mahdia movement (1881–1898) during which the country suffered one of its worst famines. There were reports that only between 35–50 percent of the population of the capital towns was able to survive the 1889–1890 famine which was known in the Sudanese tradition as the famine of *sanat sitta*. [2]

In addition to the low Nile and drought in other parts of the country, the Khalifa (or the successor of the Mahdi) had depopulated the Western parts of the country by bringing into them to the Omdurman and other military garrisons. The expansionist policies of the Khalifa were also blamed for drawing the farmers from the land to the army which contributed to the decline of crop production for so many years before the actual incidence of famine took place. The relationship between the political goals of the Mahdia during the Khalifa period and famine was described as follows:[3]

---

[1] Aspect of Sudan's Foreign Trade During the 19th Century, by Ahmed, A. H. 1974: *Sudan Notes and Records* elaborates on the expansion of Sudan trade with Egypt, Britain and Australia since 1796. Holt P. M. and Daly, M. W. 1988: A History of the Sudan, London.

[2] *Ibid.*

[3] *Ibid.*

" Besides the military and political difficulties, the Khalifa was confronted in 1889–1890 with an age-old problem, a devastating sequence of bad harvest, famine and epidemics. These natural calamities had always taxed the resources of the rulers in the Nile valley; for the Khalifa, they were devastated by his military disposition....By the total mischance, the great famine coincided with one of his major acts of policy, the enforced migration of his tribe, the Ta' aisha and their Baggara neighbours from their homelands in Darfur to Omdurman".

The 1889–1890 famine crisis revealed that man-made disasters are closely linked to the mismanagement and mal-distribution of the political and economic resources of the country. These were a result of internal political processes, but also augmented by external wars. The famine situation was so bad that food had to imported from India. This is well documented by Ahmed A. H. (1974, p. 20) who reported that,

"the intensity of the famine can be gauged by the food articles imported. Dura (sorghum) imports, mostly coming from India, exceeded the amount imported in 1889 by 30 percent, similarly flour, barley, vegetables, ghee etc., showed considerable increase".

When the famine was over, the Anglo-Egyptian troops were already at the door steps of the northern Sudanese borders with Egypt. In 1898 the battle of Omdurman had concluded a chapter of 13 years of national rule in the Sudan. The Anglo-Egyptian colonial rule continued from 1898 to 1956, a period which witnessed so many changes in the political, economic and social character of the country.

During the early period of the colonial rule many of the Western Sudanese tribes which migrated to riverian Sudan in support of the Mahdia immigrated to their home areas in Darfur and Kordofan and established large settlements around their tribal chiefs.[1] These newly created settlements were eventually used by the Anglo-Egyptian rule as a base for the local administration system. The policies of the colonial state were not accepted without resistance and several savage punitive expeditions deployed out to pacify the rebellious groups in Western and Central Sudan. So many ethnic groups, farmers and pastoralists moved into areas which were impossible for the Government troops to reach. Insecurity, inter-ethnic conflicts and the continuous fear of slave traders and fear of high taxes created considerable uncertainty among the rural population. This period was, therefore, characterized by massive population mobility. Consider for example the migration of the Nuba peoples from the tops of the Nuba Mountains to the sur-

---

[1] Mac Michael, H.A. 1912: *History of the Tribes of Northern and Central Kordofan*, Cambridge University Press and Henderson, K.D.D. 1939: The Migration of the Missiriya into south-west Kordofan, *Sudan Notes and Records*. XXII Part 1.

rounding plains,[1] the migration of the Beja to the southern parts of Kassala province to the Gash and Toker schemes[2] and the migration of the Baggara nomads from South Darfur to the whole of Kordofan and the White Nile.

However, it is not surprising that the completion of the administrative structure in the villages and for the nomadic population in 1927 had marked the expansion of the modern agricultural sector. The Gezira scheme was completed by 1926, the Nuba Mountains Cotton Industry was established in 1923 and the Toker and Gash delta schemes were operational by 1924. Later, in 1945, large-scale mechanized schemes were introduced in Gadarif area whereby large areas of land were appropriated from traditional farmers and pastoralists. It is from now on that the large-scale mechanized schemes began to proliferate in the country.[3]

The small producers were by then incorporated into the market economy producing cash crops such a cotton, sesame, groundnuts and gum Arabic for the international market and food for the emerging towns and administrative centres. This was achieved through the introduction of cash crops, and the collection of poll and crop taxes.

The growth of several urban and market centres created high demand on the surrounding forests which were used mainly for building material, firewood and charcoal. Table I shows the increase in firewood and charcoal consumption in the Sudan during the last two decades.

The situation is worse in the semi-arid zone where population growth created more demands on poor land resources to produce more food. Consequently, population mobility from one depleted area to another was common and this process continued until today. This has in the long run resulted in further degradation of the soils and the vegetation cover.[4]

---

[1] Roden, D. 1972: Down-Migration in the Nuba Hills of Southern Kordofan' in *Sudan Notes and Records* No. 53.

[2] Paul, 1954: *History of the Beja Tribes of the Sudan*, Cambridge University Press.

[3] An excellent summary of the history of the development of irrigation schemes in the Sudan is provided by Tothill, J.D. (ed), 1948: *Agriculture in the Sudan*. Oxford University Press. Osman M.S. and El Haj, H.E. 1974: Irrigation Practices and Development in the Sudan, *Sudan Notes and Records* No. 55.

[4] This process is well documented in Ibrahim, F.N. 1985: *Ecological Imbalance in the Republic of the Sudan, with special Reference to Desertification in Darfur*. Bayrueth University press. Olsson, L. 1985: *An Integrated Study of Desertification*. University of Lund, Sweden, Dept of Geography. Olsson, K. 1985: *Remote Sensing for Fuelwood Reources and Land Degradation Studies in Kordofan, the Sudan*. University of Lund, Swden, Dept of Geography. Shepherd, A. et al. 1987: *Water Planning in Arid Sudan*. Ithaca Press.

As more people began to settle, the concentration of livestock around boreholes and reservoirs contributed to lowering the carrying capacity of pasture. However, as the years passed by overstocking and over-cultivation occurred in so many areas. In other words, continuous cultivation of the same farm land, grazing in relatively smaller areas near the water sources and the cutting of trees for browsing left devastating effects on the vegetation and soil covers. The livestock situation was aggravated by relatively high growth rate which is attributed mainly to the improvement of veterinary services, the social values which encourage the ownership of large herds and the concentration of the national herd in less than 30 percent of the total grazing lands.[1]

The above features continued during independence since 1956 with a huge expansion of large-scale irrigated and rainfed mechanized schemes. The total area under mechanized rainfed cropping increased from over 2 million hectares in 1968 to about 8 million hectares in 1985/1986, with new areas in the Blue Nile, South Kordofan, the White Nile, and the northern parts of Upper Nile province. The wealthy farmers who operate under this system are allocated 1 500 feddans[2] for 25 years on leasehold basis. These large-scale private schemes swept across the traditional farms, water points grazing lands and animal routes, displacing millions of small producers. Large areas of forest were cleared and with them gum Arabic revenues, and an important source of income for the local population ceased to exist.[3]

Again, despite a steady increase in the area under cropping, the Sudan has experienced acute food shortage twice in less than two decades. These are: First, the 1967/1968 food deficit which prompted the expansion of the large-scale mechanized schemes[4]. Second, the 1983/1985 famine crisis which attracted the World's attention. Instead of the realization of some sound results from an agricultural policy which was designed to create the bread-basket of Africa and the Middle East, Sudan suffered a major setback in food production.

---

[1] Bashar, M. 1976: Problems of Range and the Settlement of Nomads in the Sudan.In Ahmed: *Some Aspects of Pastoral Nomadism in the Sudan.*Khartoum University Press.

[2] A feddan is 1.038 acres or 0.420 hectares.

[3] Adam, F.H. et al. 1983, Mechanized Agriculture in the Central Rainlands of the Sudan, in Oesterdiekhoff, P. and Wohlmuth (eds) *The Development Perspective of the Democratic Republic of the Sudan*, Weltfrum Verlag.

[4] The World Bank and the International Bank for Reconstruction and Development (IBRD)proposed an expansion of the large-scale mechanized schemes in order to be able to attain self-sufficiency from grains and to expand in the production of cash crops, especially short staple cotton and oil seeds such as sesame. Most of the farmers preferred sorghum in order to avoid the labour intensive cotton and sesame.

The extent of the ecological stress caused by the large mechanized schemes is well documented. Of more relevance to this paper is that the creation of these large-scale mechanized schemes has created at least three types of conflicts: 1) conflicts between the traditional producers and the owners of the large-scale mechanized schemes these were documented by Ahmed (1982: p.47) who argues that,[1]

> "Cultivators are forced to sell their labour cheaply, pastoral nomads are driven out of the best areas of their traditional pasture to places which not favourable to their herd growth, and agro-pastoralists are being subjected to various socio-economic pressures and forced to abandon one of the two activities and change over to agricultural labourers with low wages and lower standards of living".

This situation has also lead to another type of conflicts 2) between the local population at the vicinity of the mechanized schemes since cultivable lands are getting scarce through the years. Such conflicts were reported elsewhere[2]. These included conflicts between farmers and pastoralists as well as conflicts between those who completely lost their animal routes to the private schemes and began to move in various directions in search of new cultivable lands for themselves and grazing lands for their animals..

The most serious of these is 3) the conflict between the state policies and the interest of the small farmers and pastoralists. The state policies represented a hindrance to the small producers to immigrate into fertile lands and better pasture in the event of drought or shortage of food in their home areas. In common with Hydén, (1988:p.14),

> "the government viewed this new trend *of spontaneous settlement* with misgivings. It recognizes that development is essentially a political matter".

The spontaneous settlement has clearly signalled the irrelevance of the state to the pressing problems of the peasants and pastoralists. Furthermore, it engendered a serious conflict situation between the interest of the holders of power and their citizens.

The 1983/1985 drought revealed that popular initiatives have always been blocked by those power holders as long as such initiatives do not provide any space for them to exercise their lust for power and control.

---

[1] Ahmed, A. M. 1987: National Ambivalence and External Hegemony: The Negligence of Pastoral Nomads in the Sudan, in M. A. Mohamed Salih (ed): *Agrarian Change in the Central Rainlands, Sudan*. The Scandinavian Institute of African Studies, Sweden.

[2] Kadouf and Salih 1986:, *Land Tenure and Agricultural Development in the Rainfed Sector*, Sudan. Sudan Ministry of Planning. Mohamed Salih (ed) 1987: *Agrarian Change in the Central Rainlands, Sudan*. Scandinavian Institute of African Studies, Sweden. Mechthild, R. 1987: Land Law and Land Use Control in Western Sudan, Ithaca Press.

The very structure and location of the large-scale mechanized farms excludes and populaces. Their mere location in the intermediate land between the semi-arid zone and the rich savannah is a potential source of conflict between the pastoralists and the farmers. Furthermore it is in line either with the interest of the farmers nor with that of the pastoralists. The whole intermediate lands have now been transformed into an arena for conflict not only between the traditional producers, but also between the modern and the traditional sub-sectors of the agricultural system.

The period from 1970 to 1985 witnessed more than 20 major regional tribal conferences to solve land conflicts between various ethnic groups in the regions of Kordofan, Darfur, Bahr Al-Ghazal, Upper Nile, and the White Nile.[1]

The development of large-scale mechanized schemes has also prompted the emergence of regional political movements with clear grievances about the manner in which farming land was distributed[2]. For example, the Nuba Mountains Political Union Charter, one of the main political forces in the Nuba Mountains where about 2 million acres of land were allocated for private farmers, states that,

> "The objective of the General Union of the Nuba Mountains is to implement a land reform policy for the benefit of the indigenous farmers of the Nuba Mountains and to eradicate the feudalistic land policies and relations of production from all forms of exploitation".

This statement is endorsed by the General Union of the Nuba Mountains as a result of continuous alienation from using their lands which were distributed to private owners. As they were squeezed into a smaller and smaller area, many Nuba could not hope for any future farm expansion and some whose farm fertility has deteriorated are no longer able to open up new fertile lands. This in itself has enhanced the process of ecological degradation and prompted conflicts with the state and its declared agricultural policy. Consequently, many Nuba have joined the rank and file of the SPLA/SPLM which have been fighting since 1983 for equal rights of citizenship, equal distribution of the factors of development. The Nuba supporters of the Nuba Mountains have continuously been burning and attacking the large-mechanized schemes. Similar acts were carried out by Ingessana supporters of the SPLA/SPLM in the Blue Nile large-scale mechanized schemes. Not to

---

[1] Ahmed A. M. 1973: Nomadic Competition in the Funj area, in *Sudan Notes and Records* No.IIV and M.A. Mohamed Salih, 1979: *The Soci-Economic Basis of Intertribal Conflicts in Southern Kordofan*, MA. University of Khartoum.

[2] M. A. Mohamed Salih, 1987:*Regional Identities and the Land Question*, the Project on African Social and Cultural Identities, University of Bayrueth.

mention that large-scale mechanized schemes in Upper Nile and the Southern parts of the White Nile province ceased to operate since 1986 when the fighting between the Government troops and the SPLA/SPLM was intensified in the northern parts of the Upper Nile and Bahr Al Ghazal southern regions which were considered War Zone Number 2.

This clearly demonstrates that political conflicts created by ecological stress are not confined to the semi-arid lands. They have spread to cover a whole range of regions and ethnic groups through population mobility. Such process of the proliferation of the political conflicts can be attributed to the fact that no system of production in the modern World operates in isolation from inter-regional and inter-state activities. Furthermore, it is obvious that the conflict between the traditional and modern agricultural sub-sectors has developed into a process of economic deprivation of the rural poor who found that their very survival is continuously threatened by the dominant merchant/farmers class. This has occurred first, through the malformed market structure in which the small producers sell more and buy less and second, through the use of political coercion against the displaced and victims of ecological stress.

URBAN REFUGEES AND POLITICAL COERCION: THE 'KASHA' INSTITUTION

As has already been mentioned in the introduction to this paper, ecological degradation has forced millions of the rural poor and victims of famine to migrate to towns and centres of employment. Table II shows the number of people who were affected by famine in 1983/1985. 90 percent of those who migrated to towns, at an estimate of 4 millions, are from the semi-arid zone. The displaced and victims of famine had to make the difficult decision of leaving their home areas in the face of the fear of death from starvation.

Accordingly, the total number of the population of the Three Towns (Khartoum, Omdurman and Khartoum North) has increased from 1.15 millions in 1973 to 1.8 millions in 1983 and to over 2.5 millions in 1985/1986 following 1982–1985 drought. Although there is no reliable data to document the magnitude of rural/urban migration to other capital towns, there has been a noticeable increase in the population of some other towns such as El-Obeid, Kassala, Port Sudan, Kosti, El-Fashir, Niyala and Jenyina. These towns became major centres for relief operations.

The influx of the displaced peasants and pastoralists to towns has caused several conflicts some were manifest and others were latent. The urban population, especially in the capital towns of Khartoum,

Omdurman and Khartoum North were divided between extreme sympathy and unjustified fear of the sudden existence of hundreds of thousands of displaced peoples. The first group offered generously, while the second complainted about the pressures created by the migrants on the health, transport and education services. There was also a noticeable shortage of food as the prices began to soor. The government was anxious not to alienate the urban population and cause any undesirable political unrest. One method of achieving this is through the demolition of their houses which were built in the outskirts of town in order to force them to leave. The second method is to use the police and the army to repatriate them to their home areas.

One justification given by the Government is that the victims of famine represent potential security risk and accused them of plotting to overthrow the Government. House search and detention of some of the prominent political figure from the Nuba Mountains region were common throughout 1984/1985. The Security Police was mobilized to start a sweeping campaign known as *Kasha* to repatriate the migrants from the three towns of Khartoum, Omdurman and Khartoum North. The *Kasha* is said to be directed against the *shamasa* which can be literally translated into vagabonds or the unemployed poor classes of the urban population. The arrest and detention of the *shamasa* has always been justified by the fear that they are potential danger to security and social tranquillity in the capital towns. This for many Sudanese is a reminder of the colonial policies of 1930–1945 during and after the great depression when certain ethnic groups were repatriated to their home areas under the Town Planning Act of 1930. According to a report by Khartoum province in 1931 it is reported that every possible effort was made not only to prevent outsiders (i.e. migrants) drifting into the Three Towns in search of work, but to send unemployed persons to their villages of origin. It is abhorrant that the same process continued after independence and at a time of severe natural calamities and food shortages.

More aggression was committed by the state as the tougher Islamic laws were imposed. The Islamic *sharia* laws were implemented not to curb the widespread of corruption within the state machinery as declared by its proponents, but against the residence of the squatter settlements, the so called illicit beer brewers and the unemployed.[1] The result of the implementation of these laws is the amputation of the limps of about 200 persons, mostly accused of theft. All belong to the same group of the deprived and destitute.

---

[1] For detailed analysis of the circumstances which led to the implementation of the sharia Islamic laws and the forces behind it refer to Mansour Khalid, 1985: *Nimeiri and the Revolution of Dis-May*, London, and Abd al Rahim M. et al. (ed) 1986: *Sudan Since Independence*, Aldershot.

When famine was over, Nimeiri's regime was ousted in an uprising initiated by the *shamasa* and the urban poor amidst the declaration of lifting subsidies from bread, petrol and other essential commodities. They were soon joined by the political and educated elites and restored a democratic rule.

However, two features of political coercion which were introduced during Nimeiri's regime persisted: first, the policy of demolishing the houses of those who live in the squatter settlements. Second, the repatriation of the victims of drought to their ecologically degraded areas without any clear policy as to how they should survive when they are there once again.

These two features of repression suggest that the repressive policies of the Sudanese state towards the poor and the victims of ecological stress have persisted. They seem to be having the same manifestations under the dictatorship of Nimeiri and the democratically elected Government of Saddig Al-Mahdi. The only difference is that there is more room for political debate among the educated and political elites than during Nimeiri's regime. This suggests that the powerful political interest groups are capable of manipulating the state apparatus, regardless of its nature (totalitarian or democratic), to protect their interest. The persistence of the *Kasha* institution is a vivid example of one of the methods of political coercion exercised by the state to protect the interest of such groups.

CONCLUSIONS

During the peak of the famine crisis in 1985, the Sudan Government was forced by political unrest and food riots to declare the Sudan a disaster country. This was largely mitigated by the poor performance and discouraging economic prospect of the country. The realization that the state resources were so limited to cope with the situation can be spelt out by providing a brief review of the economic situation. The gross national per capita (GNP) was estimated at 400 US dollars. Since 1977/1982 the Gross Domestic Product (GDP) increased only by 2.6 percent with a population growth rate of 2.8 percent. It is observed that the average per head income in 1985 was 12 percent lower than that of 1971. The decline in the economy is attributed largely to three years of drought (19982/1985) which ruined Sudanese agriculture. Debt was about 10 billion US dollars with an annual service repayment equal to the total foreign exchange earnings. This was at the time when food deficit during 1984/1985 was estimated at 60% (i.e. 1.6 million tons of grain alone).[1] It became clear in the circumstances that the meagre financial resources of the country were absolutely incapable of bridging the food deficit.

The economic crisis is summarized by Ali A.A. (1985: pp.9–16)

> "i—Slow growth, ii—worsening balance of payment, iii—deteriorating terms of trade, iv—slow growing exports, v—mounting debts, vi—worsening budget balance and vii—excessive monetary expansion."[2]

It is shocking that Sudan Government, at the time, was exporting large quantities of sorghum, the staple food for the majority of the population, in 1983, 1984 and 1985, amidst indications that production was deteriorating (Table III). The Government was therefore forced to boost its exports, even if that meant the exportation of food during a period of severe hardship to finance its mounting debt and to improve its continuous deficit in the balance of payments. The situation, however, resembles a picture of the disjunction between the state and the civil society in Africa as depicted by Hydén (1988: p.12). He argues that,

> "The rural people were no longer ready to sustain their belief that Governments could continue to serve as the sole *and trusted* dispenser of resources and benefits. Feeling deceived, it is only natural that people in the rural areas *as well as the urban centres* turned towards things proximate and known. Faced with declining public services, people began to explore alternatives "[3]

Again the initiatives of the rural people were intercepted by the state machinery which prevented them from securing their survival by moving into new lands. Ecological stress and the growing recognition of its consequences, therefore, are inseparable part of the formal political activity of the African state today. Reports of food riots culminating from ecological stress or the distributive mechanisms of the state were common in Africa during the first half of the 1980s. Consider for example the food riots in Sudan, Nigeria, Zaire, Egypt, and Algeria. What is common in all these cases is that the nature and magnitude of the conflicts are mediated through the interaction between various interest groups and their ability to use the pervasiveness of the state apparatus. The Sudanese experience has shown that a state under external (financial constraints, debt burden, unequal exchange, etc.) and internal (underdevelopment, economic crisis, drought, famine, etc.) pressures is incapable of properly and competently solving the prob-

---

[1] International Labour Office, 1985: *After the Famine*, Geneva.

[2] Ali, A. A. 1985: *The Sudan Economy in Disarray*, Ithaca Press.

[3] Hydén, G. 1988: State and Nation Under Stress, a paper presented in the Symposium on Swedish Developement Co-operation with sub-Saharan Africa in the 1990s, Saltsjöbaden, Sweden.

lems emanating from ecological stress. Hence, the state's effort to deal with development have been diverted to one of survival management. Since survival management requires certain structural and infrastructure amenities, which most African states are lacking, the general tendency has so far been the deployment of coercive measures to curtail the undesirable consequences of ecological stress. However, two limits are involved: first, there are limits to the coercive nature of the state apparatus which could instigate popular uprising. A clear evidence of this is the increasing defiance of the legitimacy of the state through food riots during drought and famine. Second, the under-development situation in which most of the African states exist limits their respective solutions to the use of coercion to protect their own survival. This sufficiently signals the perception gap between the populace and the government which could also explain the indifference of the state to self-perpetuated settlement initiatives during drought. Since the control of landed resources is one of the main domains in which the state exercises its jurisdiction powers, the link between ecological stress and political conflicts is a forcible expression of the land-based politics in the dominantly agricultural African states. The relationship between ecological stress and political conflicts could therefore be a result of ecological or political imbalances which, in both cases, underline the interdependence between ecology and the prevailing political activities.

TABLES

Table I. *Firewood and charcoal consumption in the Sudan 1962–1985 (in 1000 cubic metres)*

Fuel type	1962	1975	1980	estimates for 1985
Firewood	12,048	23,680	23,490	30,000
Charcoal	8,362	11,240	44,082	50,000
Total	20,410	34,920	67,572	80, 000

Source:Forestry Reports 1962/1985 quoted in Awouda, E.M.1986

Table II. *Drought victims and displaced persons by region.*

Regions	Drought victims		Displaced persons	
	No.	%	No.	%
Darfur	2,830,000	33.7	500,000	27.8
Kordofan	2,870,000	4.2	425,000	23.7
Central	1,350,000	16.0	420,000	23.4
Northern	150,000	1.8	50,000	2.8
Eastern	2,200,000	14.3	400,000	22.3
TOTAL	8,400,000	100.0	1,795,000	100.0

Source: After the Famine, International Labour office, Geneva, Sept. 1985: p. 19.

Table III. *Sudan exports of sorghum 1981–1985*
(in 1000 metric tons)

Year	Quantity	Value in LS	Exports
1981	241.3	42.9	12.0
1982	412.8	107.5	22.3
1984	24.9	7.2	13.9
1985	----	----	13.1

Source: Economic Survey, Economic 1986/1987 Research Administration, Ministry of finance and Economic Planning, Khartoum, tables 5/5 and 4/6, pp. 205–206.

Land Degradation and Class Struggle in Rural Lesotho

Kwesi K. Prah

"Hele! Where has this *Lengope* (sloot, watercourse) come from? It was not here when I was a boy. The ground was flat and level, where now I see a deep and yawning chasm. Hele!..." (John Widdicome *In the Lesuto* , 1985)

INTRODUCTION

Lesotho is often euphemistically called "the Switzerland of Africa", "Kingdom of the sky", "Roof of Africa" and "Mountain Kingdom". These descriptions variously allude to its mountainous topography, the beautiful and uniquely rugged scenery, the dramatic elevations in the landscape and the healthy temperate climate. The land is largely bald, and as one early writer has put it "almost entirely destitute of trees, save at the various magistracies and mission and trading stations, and at some of the larger and better-class villages".[1] The population, estimated at 1.61 million for 1987 is supported on an arable acreage of less than a million acres.[2] Crop yields are very low, and there is increasing dependence on imported food and international food aid.[3] Much of the grazing land is stressed through overstocking and the quality of stock is much less than desirable. Remarkably, the Lesotho eco-system had a mere century ago supported a wide variety of fauna which have since become extinct in the country.[4] For a country which in 1863 was described as "the granary of the Free State and part of the (Cape) Colony", present conditions represent in historical perspective, extreme retrogression in agricultural production.[5] Significantly, the decline of agricultural production in Lesotho has gone on hand in hand with the phenomenon of migrant labour to South Africa. From the 1870s onwards, the Bashoto were drawn into mine labour, attracted by the wages and economic forces of the emerging mining capital of South Africa. Ashton has estimated that by 1875 out of a population of 127,325 about 15,000 men were getting passes to seek employment in South Africa for varying contract periods. By 1884 the number had doubled.[6] 20,00 labour passes were in use by 1892.[7] The Colonial Office Annual Report for 1906–07 recognized that "the native population is showing a tendency to divide itself into two distinct classes of labour, namely the agriculturalist and the mine labourer".[8] While agricultura products such as wool and grain continued to feature fairly prominently during most of the first two decades of this century, agricultural production started to decline significantly from 1928. By the 1930s, about 50 per cent of the able-bodied male population had been proletarianized as mine migrant labourers in South Africa. The steady growth of mining and, subsequently, manufacturing capital in South Africa accelerated the process of attracting labour throughout the southern African sub-region. Through a plethora of legislations affecting "natives" it was assured that African labour would be tied to "Native Reserves" where labour would be allowed to reproduce itself as cheaply as possible, and hence ensuring the maintenance of a "reserve army of labour". Geographically buried in the heart of South Africa, Lesotho was together with Botswana, Swaziland, Malawi, the Rhodesians, and Mozambique subject to the proletarianizing influences of mining and manufacturing capital concentrated in South Africa. Within South Africa, the economists of race and class ensured that development of infrastructure on the basis of racial criteria, tariffs against non.South African agricultural produce, the subsidization of the white enclave of capitalist farming within South Africa, and taxation of the African agricultural commodity producer, catalysed the process of slowly eliminating the African rural producer as a self-sufficient economic and social category. This also meant that within the rural "native reserves", to which category for all practical economic purposes Lesotho belonged, the primary means of production land, was in value undermined by the emergent patterns of South African racist capitalism. As a viable means of production, it declined in significance and increasingly remained a feature of pre-capitalist social relations of production. Its quality in turn declined, becoming a casualty of ever-increasing soil erosion, overstocking and range mis-management, and general soil-exhaustion. With little or no capital infusion to improve the quality of land, Lesotho, once home of some of the best soils in Southern Africa has become a country in which;

> "the most vivid memory retained by the visitor who loves the land is of creeping tentacles of the erosion gulleys, eating into the lowland soils like some malignant canker".[9]

In a more recent report, the Central Planning and Development Office, Government of Lesotho, admitted tersely, the gravity of the situation:

> "The devastating extent to which Lesotho's soil is being eroded cannot be too greatly emphasized. Gulley formations in inadequately vegetated slopes are evident everywhere and particularly in the lowlands. Widespread sheet erosion, though less visible in effect, is destroying the fertility of both range and

cultivated land. These disastrous processes are encouraged by Lesotho's rugged topography, extreme seasonal changes i climates, and highly erodible soils in the lowlands. The condition of the land has been further aggravated by poor cropping methods cultivation of too steep and inadequately protected land, neglected from earlier conservation measures, and especially by overgrazing.[10]

Indeed the challenges of soil conservation and "donga" control in Lesotho are vast. Anti-erosion activity started in earnest in 1933 when initial contour banks and buffer strips were first constructed, 188,000 acres of land had been designated for buffer strips, over 6,000 kilometres of diversion furrow and 787 dams had been constructed. All this construction however, was already then "a holding operation, designed to delay and reduce erosion rather than stop this fearful force".[11] Conditions have continued to deteriorate throughout the 1970s and 1980s.[12] Thus although since the Sir Alan Pim Report of 1935, official mind and activity has been brought to bear on the problem of environmental degradation through soil erosion, comprehensive and effective conservation has practically eluded both the colonial and post-colonial authorities in Lesotho during a period of over a half-century of official concern.[13] An important factor in explaining the lapse and ineffectiveness of the colonial and post-colonial authorities in dealing with the problems of land degradation is that approaches to solving the problems underestimate social factors. This point is, however, somewhat understood in an internal discussion paper produced for the Land Conservation and Range Management Project of the Government of Lesotho, a decade ago.[14] The paper observed that the relevance of social factors for soil conservation and range management in Lesotho "has emerged from successive analyses of the problems of conservation and range management... one might go so far as to say that conservation and range management in Lesotho, are essentially social problems"[15] It is further revealed that a recurring feature of the historical trajectory of earlier efforts at conservation and range management in both the colonial and post-colonial periods has been the divergence of opinion between official government supported planners and the peasantry. While planners have continued to offer "a large range of technical solutions to the problem" it is clear from evidence arising out of various surveys that;

> "... villagers do not see conservation as a priority area. There are quite rational reasons why this is the case... villagers' relative lack of concern with problems of erosion appears to be quite a rational choice in the face of the more immediate problems facing the farmers, and in the context of the real options in the migrant reserve economy"[16]

The recognition of the socio-logical and political implications of the problem of the land ownership and land use need further and closer dissection. They appear to be influenced primarily by class factors operating within the social structure of Lesotho, and Lesotho's position within the wider context of the political economy of South Africa and the development of capitalism in the socio-economic orbit of South Africa.

CLASS, PRE-CAPITALISM AND CAPITALISM

If the class concept is understood to designate a group, which within a given social system of production relations, with particular respect to the ownership of the means of production, the patterns of distribution and exchange, have essentially the same position, it is arguable that the overwhelming majority of the able-bodied male population of Lesotho today is largely proletarianized through the migratory labour system. They are, however, not engaged and proletarized within the economy and polity of Lesotho treated as an integer system, but rather, they have been proletarianized by South African based capital concentrated in mining. Since 1982–83, over 50 per cent of the Gross National Product arises out of earnings, remittances and deferred payments of migrant labourers working in South Africa. Migrant labour income exceeds actual Gross Domestic Product.[17] Baring the possibility of dramatic repatriation of Lesotho migrant labourers from South Africa, projections indicated that the proletarianization of the Basotho reserve army of labour is likely to increase.[18]

The absence of able-bodied males in rural Lesotho has a detrimental effect on overall agricultural productivity. Most of the heavy rural farm work is done by women and children. The yield per hectare of key grains has over the past two decades been in steady decline. Income of migrant workers is often used to purchase livestock (of relatively poor quality) rather than improving crop production methods, and innovative inputs. This puts an increasing strain on the carrying capacity of rural land and accelerates the processes of soil erosion and land degradation.[19] The salary and wage differentials between employment possibilities in South Africa and Lesotho cause a skills drain.[20] The social toll of the system on family life in rural Lesotho is debilitating and irregular, with children and wives seeing respectively their fathers and husbands only sporadically.[21]

While the proletarianization of the Basotho labour force is growing, the phenomenon of rural landlessness is also growing. Agricultural censuses over the past two decades indicate that rural Lesotho is now overpopulated on the basis of the criterium of allocation of land. Between 1969/70 and 1979/80, the number of rural landless households rose from 26,919 to 45,549 or from 12.8 per cent to 20.7 per cent of all rural households.[22] It has been estimated that already the number of grazing animals in rural Lesotho is 300 per cent above the land's carrying capacity. The consequent overgrazing and poor conservation measures have resulted in the further shrinking of the rural productive capacity. Arable land is seriously constrained, and is down to 12% of the country's total land area.[23]

The state sector of the economy is growing rapidly and drawing heavily on total expenditure. Thus it is arguable that the labour of miners is largely supporting the growth of a burgeoning bureaucracy and other parts of the state apparatus. The petty bureaucracy is increasingly characterized by underemployment and the ravages of Parkinson's law.

There is hardly any growth in the accumulative capacity or process of the rudimentary entrepreneurial class. "Penny capitalism" is still predominant and thus most of the earnings of foreign exchange acquisition i the country flows into South Africa.

In the rural areas, pre-capitalist land and social relations are still operative, and this arrests the process of accumulation and the development of the productive forces. It reinforces the maintenance of a backward rural political–economy and enhances the preservation of Lesotho as a labour reserve for the South African capital–labour market. In an interesting paper entitled; Land Reform Patterns in Africa and their Implications for Agricultural Development: The Case of Botswana, Lesotho and Swaziland (1975), Sefali argued that "communal land is not only an economic category, but also a foundation of the socio-cultural fabric of the nation. Hence its reform might have a disruptive effect on the whole fabric of the nation-building process". this argument is in substance populist nostalgia. The basis of communal land rights in Lesotho has been irreparably eroded over the past three decades. Rural landlessness and inequalities are the substantial realities of the contemporary situation. The proletarianization of rural labour into mine labour working in South Africa is a highly advanced process, with about three-quarters of the able-bodied male population engaged for extended periods as mine labour in South Africa. What appears to be important, is the development if the productive forces. While cooperativisation on a wide sociological scale is in the long run desireable, it is doubtful whether crucial historical processes can be "jumped".

The Basotho during the late pre-colonial period had developed a pre-capitalist social formation based on rank and class differentiation. The late precolonial polity would fall under the category of social organization, described by Fortes and Pritchard as "State societies".[24] In as far as the dominant Bakwena clan extracted social surpluses from the commoners, institutional structures existed for this process.[25]

Institutions like the *Tsimo ea lira* (land from which the material results of production are left to the disposal of the chief), *Malahleha* cattle (lost cattle) specific forms of *Letsema* (collective labour) and various forms of tribute, were channels for the extraction of social surplus. Vassalage in forms like *Letona* and *Lefielo* underpinned the recognition of high rank and was rewarded. Depressed classes like the *Mekhoba* and *Mohlanka* provided serf-like labour at the base of society.

There is a record of Mofolo complaining that:

> ... a native miller is required to close his mill and go to the chiefs' *letsema* to plough, five or six days a year, according to the number of the chiefs' *masimo* (fields). Another five or six days for scoffling, and there is the reaping, the gathering in of the *mabele* and the mealies, and *ho pola*. In addition... the chief can send him wherever and whenever he likes, it matters not to the chief whether the poor miller is away a month or two or six from his business. So that really the poor man has no time to devote to his business. This applies to all men whom he may employ.[26]

Kimble indicates that during the early decades of colonialism, in response to pressures of the evolving political-economy, the chiefs started intensifying their extraction of social surplus and consolidate their primacy in the land tenure system. *Matsema* obligations were increased, and the chiefs started treating the land under their control "as if they had rights to private or permanent ownership over it". The Paramount Chief began to define for himself exclusive control over the pasture land in the *Maluti*. Also the chiefs asserted their duties in tax collection since this reflected directly on their percentage takings, for the higher the taxes collected, the greater the percentage paid to the chief.[27]

The land tenure system inherited from the pre-colonial period has been preserved in its essential form until the Land Act of 1979. Basically the old tenure system held that all land was vested in the trusteeship of the King, who held this trust on behalf of the Basotho. The King in turn delegated the practicalities of allocation and apportionment to the chiefs and headmen. According to Mosaase, "the Kwena lineage has ruling dynasty of more than 1086 chiefs who have power to allocate land".[28] The hierarchy of chieftainship in Lesotho operates as follows:

Hierarchy of chieftainship [29]

Area of responsibility	Title	No. of officers
Nation	King	1
Ward	Principal Chief	22
Village Area (Group)	Ordinary Chief and Senior Headmen	1,200
Village	Village Head	5,000

Under customary land law freeehold land did not exist. Land of the deceased eventually reverted back to the jurisdiction of the chief for reallocation. This usage has in modern times tended to undervalue land, since only usufruct rights are permitted. Contemporarily the average size of farm holding is 5 acres. Less than 3 per cent of the farm holdings are about 15 acres or more.[30] Qualification for land allocation has been tied to marriage in the natural expectation of social reproduction and family life.

The Coutts Report of 1966 drew attention to the fact that "until some alterations are made in the present system of land tenure, there seems little hope of real advance in the development of Lesotho".[31] It further emphasized there was need for the institution of leaseholding in land relations.[32]

In 1973, the National Interim Assembly promulgated laws in order to control land allocation and bring efficiency into land administration. The two laws were the Land Act 1973, and the Administration of Land Act 1973. The Land Act of 1973 was repealed in 1980 through the 1979 Land Act. The administration of Land Act of 1973 which introduced the concept of leasehold never became operational on account of the strong opposition from the chiefly groups. The Administration of Land Act had theoretically repealed the tradition-bound Land Procedure Act of 1967. The opposition and obstacles to change in land tenure was such that a chaotic land tenure conjuncture arose in the early 1970s. This was a reflection of the contending chiefly interests and the emergent middling classes.

In 1979 the new Land Act was enacted. This Act still recoginzed the principle of the King being guardian of the nation's land. But in essence the new Act attempted to supersede the constraints of pre-capitalist relations in land through a clear introduction of the leasehold concept.

While the 1979 Act recognized the negotiability and transferability of land on leasehold terms, it stopped short of freehold tenure. In this sense the 1979 Act can be regarded as a compromise measure between the chiefly and feudal interests in the country and the pressures for capitalist penetration into agriculture and rural production relations. Nevertheless, what has in hindsight emerged is that as one report has noted (1986–87) "the new government has already acknowledged the failure to implement the 1979 Land Tenure Act, and is seriously considering the necessary policy changes to carry through what needs to be done".[33] In 1987, a new land commission was instituted. Its findings are a matter of keen interest to Bashotho and other interested observers. The 1987 Land Policy Review Commission was established in February of that year. It was mandated to scrutinize and review land tenure arrangements and their administration within the framework of existing legislation. The Commission was expected to advise government on any changes necessary for the proper and efficient administration of land in Lesotho. The following are the essential and most pressing problems as identified by the commission:

(a) The misuse and mismanagement of arable and grazing lands throughout the country and lack of motivation to invest in land improvement and soil conservation.

(b) The failure of the land Committee system in rural areas, and the reluctance of chiefs to cooperate in implementing land laws. This results in illegal allocations of land.

(c) Total lack of control on urban land use as a result of no clear housing policy and urban development strategy.

(d) Lack of qualified manpower, disciplined personnel, finance and political will to implement approved land policy. This results in inefficiency of administrative land institutions which causes hardships to members of the public.

(e) The concern by many Basotho that land is increasingly passing from the hands of the ordinary unsophisticated Basotho into the hands of the rich, elites and other aliens who do not necessarily invest the country.

(f) A highly centralised land administration system which results in bottlenecks, bureaucracy and corruption. [34]

One of the fundamental criticism of the commission was with regards to the main principle of land tenure in Lesotho. The commission argued that the key principle that all land in Lesotho is vested in the Basotho nation is accepted by all citizens of the country. The Land Act of 1979 accepts this principle but "unlike previous land laws it mentions that the nation's asset is held in trust by the state instead of His Majesty the King". This was a source of discontent to many Basotho because the Sesotho translation equated state with government. "The Commission is aware that the word State in the Act is used in a broad sense to include the Head of State and the whole machinery which runs the country and the State does not change while government can change. However, to dispel any misconceptions the Commission recommends that the Law should be amended to stress that the land is held in trust by His Majesty the King as Head of State".[35]

The land policy Review Commission of 1987 made further recommendations with regards to; residential sites in rural areas; commercial and industrial land use in rural areas; uncontrolled encroachments on farm lands; measures to improve agricultural production; re-grouping of villages for administrative purposes; specific land use plans; the need to encourage farmers to specialise in farming; issues of management and protection of range lands especially the need to introduce grazing fees; a better conceived urban development strategy; the uncertainties of land allocations in urban areas; etc.[36]

In on-going research undertaken by Volker Rein in Ha Sekake, Qhacha's Nek district, the divergence, and conflict in socio-economic interests of the Chieftainess who commands 20 acres for herself as against the more average 1.5 to 3 acres held by the hoi polloi, is revealing 55 per cent dispose over fields (most of the households dispose of cultivatible land from 1.5 to 3 acres), 13 per cent do not have even a garden.[37] Many respondents suggested that "the Chieftainess does not allocate land properly". 53 per cent of Rein's sample favoured the institution of a land right which legalizes a universal disposal of land whether under cultivation or not. Such an arrangement it was suggested, should include sale and inheritance by both sexes. A commonly held charge against the Chieftainess was that she abused her rights on land allocation for monetary gain.[38] She is not unique in rural Lesotho. A recently published government report indicates that illegal allocations are effected by backdating of title documents to land by local chiefs to before 16th June 1980, the date of commencement of the operation of the Land Act 1979. Factors contributory to illegal allocations are many, but the major reasons are according to the report to be the following.

1. There has been opposition by the chiefs to the Land Act 1979, most probably because the Land Act 1979 was viewed as weakening the chiefs powers over their subjects by removing the very cornerstone; that of land allocation.
2. There appears to be a fear prevalent among some landholders, especially fieldowners, generally described as "progressive farmers" in Lesotho that if and when their lands are required for public purposes there would be scanty or no compensation.
3. Inadequate security of title to agricultural lands near urban areas is being disposed of for fear of inadequate or no compensation in case they are acquired by the authorities for public purposes.[39]

The "progressive farmers" in essence represent the emergent middling social category in the rural areas. While many are urban-based, they are displaying increasing interest in rural entrepreneurial activity. Their material interests diverge from those of the traditional chiefly elements. 28 per cent of Rein's rural sample preferred the current usufructuary land right. The predominant reason for this is its customary status. "Fields can be confiscated from lazy people and given to landless villages" it was suggested by some. "European land right confuses the people and it is unjust (rich people became richer" it is said by others. 15 per cent of the respondents proposed a mixture of land tenure approaches.[40] Such views underscore the phenomenon of class contradiction. The 1979 Land Act is in implementation hindered by the dominant classes, who ultimately regard the customary pre-capitalist land teneurial relations as the cornerstone of their social power. They are well-represented in the existing state.

CONCLUSION

The traditional land tenure system in Lesotho is often used by the pro-chiefly lobby to argue that it provides a social insurance to returning migrants. While to an increasingly limited extent it provides this, in essence it constitutes a principal pillar in the maintenance of pre-capitalist social relations in rural Lesotho. Its very nature hinders the process of infusion of capital on an individual basis, and thus under-values rural land. Because rural land is under-capitalized it is highly susceptible to environmental degradation and neglect.

The migrant labour system thrives on a backward agricultural system, and pre-capitalist residues of social relations fetter the development of productive forces. It is arguable that the type of "natural" economy based on extended consanguinity and a *gemeinschaft* ethos is not sufficiently dynamic towards stimulating accumulation. This is because the individual producer situated within a *gemeinschaft* oriented society has poor incentives towards undertaking additional labour. Furthermore, under such conditions surpluses tend to by disbursed extendedly and consumed. The penetrations of the cash nexus and rise of increasing monetary requirements promotes migration to areas of higher labour marketability which would provide the necessary and desired monetary rewards. Thus the rationale and nostalgia for social security based on pre-capitalist assumptions stands in contradiction with the historical terms of capitalist penetration and a modern peasant economy.

In the closely-knit society that Lesotho is in general, and rural Lesotho is in particular, social and *gemeinschaft* power structures are structured invariably on kinship, consanguinal, and social obligations. Hitherto, migrants and rural producers have not been able to organize and develop structures representative of their interests. Circumstances commit them to traditional rulers and leaders in the expression of their class interests. Thus a system of exploitation by the state apparatus and its bureaucracy ensures that their access to capital and infrastructural resources is hindered. Rather, these resources are directed towards the subsidization of the culture of consumption of the dominant groups in the state structure.

Migrant labour does not favour attention to the quality of agricultural land. It removes from rural production its key factor of production, i.e. labour, and exposes the means of production to a process of marginalization and retrogression in its productive capacity.

The contending interests of the emergent middling petty bourgeois classes on one hand and the chiefly feudal classes are pulling at different ends of the 1979 Land Act. This has so far in effect paralysed that fuller implementation of the Act. If rural development is to advance in Lesotho, land relations would have to be freed from pre-capitalist tenurial relations.

NOTES

1. Minnie Martin 1903: *Basutoland, Its Legends and Customs*. London. Quoted here from the 1969 edition. New York, p. 6.
2. See *Country Profile. Botswana. Lesotho. Swaziland*. The Economist Intelligence Unit, 1988-89. London 1988.
3. Atnafu Tola. Food Security in Lesotho. The Challenge and the Strategy In Kwesi K. Prah (ed.) *Food Security Issues in Southern Africa*. ISAS. Southern African Studies Series. No. 4. 1988. In 1983–84 Lesotho met only 40 percent of its total supply from domestic production The largest amount of food, 46 percent, came from commercial imports. 14 percent was covered by food aid, mainly coming from Canada, Belgium, China, Germany, Italy, USA. and the EEC. The USA is the largest donor accounting for about 60 percent of the food aid to Lesotho with EEC contributing 36 percent. *Ibid.*
4. See John Widdicombe *In the Lesuto*. London 1895 pp 11-12. Writing in 1895 Widdicombe remarks that; "fifty years ago Basutoland was full of wild animals The lion roamed... Troops of quaggas... elands, springboks, blesboks, reeboks, rietboks... hyaenas, panthers, ounces, jackals, wolves, baboons, and wild dogs abounded. Almost all these are now gone".
5. Robert C Germond, 1967: *Chronicles of Basutoland*. Morija. p 459.
6. E. H. Ashton, 1962: *The Basuto* . London. Quoted here from Colin Murray. *Families Divided* Jo'burg p 12.
7. Colin Murray. *Ibid.*
8. Quoted here from C Murray *Ibid.*
9. Lesotho's Grim Fight Against Erosion *Lesotho Quarterly*, Vol 2, January 1968. It is two decades ago that these observations were made The writer had then observed that "the challenge faces the nation is to stop the frightening destruction of soil and grass, to improve the fertility and management of arable lands and natural grazing;

and to do whatever may be possible to reclaim the erosion gulleys". Twenty years later the Problems have growth further in magnitude.

10. Lesotho *Donor Conference Papers* . CPOO. Maseru. 1977. p. AG-2S.

11. *Ibid*. Additionally. it is observed that a conservation programme cannot succeed unless an allied programme is undertaken to educate those who control and use the land in the techniques of conservation farming. While this is true, more importantly, the value of land as means of Production would need optimization to make local agriculture as a production system competitive and viable.

12. See Bernt Rydgren, 1985: *Land use and Soil Erosion in the Maphutseng Soil Conservation Area.* Uppsala University Geography Department Publication. p. 1.

13. See Sir Alan Pim, *Financial and Economic Position of Basutoland.* HMSO. London. 1935. See also Qalabane K.Chakela, *Water and Soil Resources of Lesotho A Review and Bibliography.* Publication of the Swedish Secretariat for International Ecology. 1973 pp. 3-6. An early statement on the problems of ecological stress is provided by.R R.Staples and W K Hudson *An Ecological Survey of the Mountain Areas of Basutoland.* London 1938.

14. Land Conservation and Range Management Project Paper (LC/RM PP). Internal Discussion Paper. A Proposal for the Conceptual Design of the Land Conservation and Range Management Project. 1972.

15. *Ibid*. p. l

16. *Ibid*.

17. Final Report Assistance to Lesotho on the Mining Repatriation Issues. Development Associates Inc. Arlington 1987.

18. *Ibid*. Emerging political and economic trends suggest that South Africa's reliance of Basotho labour is decreasing. This in effect will put pressure of rural Lesotho resources for Providing productive settings. This will also increase the pressure on land and an already stressed environment. See p. 61. Ibid. See also Meredith B Burke. *The Outlook for Labor Force Growth and Employment in Lesotho 1980-2000.* World Bank and UNDP Team CPDO. Maseru, July 1981.

19. See Lesotho *Donor Conference Papers.* CPDO. Maseru, 1977, p ME-3.

20. See K K. Prah, (Forthcoming) *The Bantustan Brain Gain.* ISAS Southern African Studies Series. No 5 1988.

21. See C Murray's excellent study, *Families Oivided. Op. cit* .

22. *Ibid* p 37–38. See also Ministry of Agriculture *Agriculture Development and Employment Issues in Lesotho.* In JASPA 1983, p. 111 Table 1 and Bureau of Statistics, Agricultural Census of 1969–70 and 1979–80. In 1983. official government ideology was suggesting that "...population pressures and economic growth have introduced major distortions in traditional utilization of land and livestock. Consequently, land distribution has become increasingly unequal" (Lesotho Country Report on *Progress in Agrarian Reform and Rural Development.* Interministerial Task Force Government of Lesotho. Maseru, March 1983). Hard evidence would however suggest that rather it is more the misallocation and appropriation by the chiefly classes which has provided the greatest impetus to rural inequalities in landholding. In as far as the Lekhanya Administration does not represent a fundamental overthrow of the ancien regime, the development of rural inequalities in landholding continues under the pressure of essentially the same social forces.

23 *Ibid*. p. 2.

24. Meyer Fortes and E Evans-Pritchard *African Political Systems.* London 1940. I have drawn attention to the inadequacies of this theory in *Current Anthropology in Africa. A Critique of Functionalist Ahistoricism.* Occasional Paper No 3. Institute of Southern African Studies, National University of Lesotho 1986.

25. While it will be wrong to suggest that the Bakwena constituted exclusively the ruling group, it is possible to say that they dominated the pre-colonial Polity clans like the *Bataung, Basia. Baphuthing, Batloung,* and especially the *Bafokeng,* who had been earlier Bantu-speaking elements who penetrated the area were reduced to some measure of subsidiariness to the pre-eminent Bakwena See also W J Breytenbach, *Crocodiles and Commoners in Lesotho.* Communications of the Africa Institute,No. 4. Pretoria.

26. Thomas Mofolo to Resident Commissioner. In S3/26/12/10 Maseru Archives quoted here from J. Kimble. Aspects of the Penetration of Capitalism into Colonial Basutoland. 1890-1930 In *Class Formation and Class Struggle.* Proceedings 4th Annual SAUSSC Conference 1981.

27. J. Kimble. *Ibid.*

28 A. Mosaase, 1983: An Analysis of Existing Land Tenure in Lesotho and Experience in the Implementation of Current Land Policy. Mimeo. Commonwealth Association of Surveyors and Land Economists. Harare .

29. C.G. Wenner,1982: *Soil Conservation in Lesotho .* p. 17.

30. C.G. Wenner, *Ibid.*

31. Sir Walter Coutt, 1966: *Report on the Structure and Administration of the Lesotho Government* Maseru, p. 23.

32. *Ibid.* The need for revision of the Lesotho Land Tenure System had been a recurring theme over the past three decades. See I. Hamnett, Some Problems in the Assessment of Land Shortage. A Case Study in Lesotho *African Affairs* Vol. 72 1973. V. Sheddick, *Land Tenure in Basutoland.* HMSO London. 1954. C Williams, *Lesotho Land Tenure and Economic Development.* Pretoria 1972, Review of Lesotho Second Five-Year Development Plan. CPDO, March 1975

33. Final Report Assistance to Lesotho on the Mining Repatriation Issues. Development Associates Inc. Arlington 1987.

34. Summary of Recommendations. Land Policy Review Commission. Mimeo. Maseru. 1987.

35. *Ibid.*

36. *Ibid.*

37. Volker Rein, 1988: "Indigenous Development Conceptions and Cooperative Forms of Self-Help in a rural area of Lesotho: A Case Study of Ha Sekake ". Mimeo.

38. *Ibid.*

39. See Maseru Development Plan, Working Paper No 4. Land Tenure. Physical Planning Division, Dept. of Lands Surveys and Physical Planning.

40. Volker Rein. *Op cit.*

Political Instability and Ecological Stress in Eastern Africa

W.P. Ezaza and Haroub Othman

INTRODUCTION

Ecological stress and political conflict can be considered as two sides of the same coin.

The relationship between ecological stress and political conflict has already been clearly demonstrated by the events which took place in Africa in the early 1970's and 1980's. Over 30 million people from nearly 20 independent African states were suddenly hit by what at first appeared as drought (Timberlake, 1986).

Many thousands more were forced to evacuate their homes with their families and livestock to search for food, pasture, water and peace across their regional and national borders.

In northeastern Africa and in the Sahel Region, the failure of the land to feed its own people led to social unrest and frustration which triggered off massive migrations across the borders.

Similarly, political upheavals in eastern and southern Africa have generated social insecurity, loss of lives and property, mismanagement of the available resources and led to collapse of productive systems.

While the causes of the ecological stress and political instability in the Horn of Africa and in the Sahelian Region were connected with environmental bankruptcy (Timberlake, 1986), those in eastern and southern Africa (Mozambique, Angola, Botswana and Namibia), were designed and engineered by the racist South African government (Othman and Kiss, 1988).

In Mozambique, Uganda and Rwanda, political upheavals have been shaking these countries since their independence in 1960s. As a result, social systems and economic activities have over these years been disrupted thus causing great resource deterioration.

Since conflict and resource degradation intensify each other, these conflicts have undermined the very life support system upon which the livelihood of those affected depend. In Uganda for instance, political instability led once to a total breakdown of economic activities. In Mozambique, it initiated social insecurity, forcing thousands of innocent people to flee their land and squat on the borders of Malawi, Zimbabwe and Tanzania.

Thus the overall economic and biosocial consequences of political insecurity and ecological stress are taking many people of eastern and southern Africa unawares faster than they can adjust themselves to the deteriorating conditions.

Although a causal relationship between political conflicts and ecological stress may appear obvious and simple to establish, the reverse is surprisingly even more difficult.

This is partly true because unlike the former, the ecological processes which eventually culminate into ecological imbalances and create political conflict happen to manifest the end results of a prolonged environmental deterioration.

This paper attempts to elucidate the obvious relationship between political instability and ecological stress in eastern Africa through examining the cases of Uganda and Mozambique.

POLITICAL CONFLICT, SOCIAL DESTABILIZATION AND ECOLOGICAL STRESS IN UGANDA

The political upheavals which have shaken Uganda for 20 years now have destabilized the social security, disrupted the economic activity and caused severe resource mismanagement and degradation.

It is not clear whether it is a result of ecological stress which has caused the upheavals but the fact that the power struggles were mainly aimed at access to and control of the vital resources in Uganda cannot be denied.

For example, the expulsion of the Kabaka of the Baganda Kingdom in 1966 was not only that Milton Obote—the then Prime Minister—disliked the existence of a 'Buganda State' within the State of Uganda but he also wanted access and control of the vital economic resources, most of which are located in the Buganda Region.

Also the overthrow of Obote by Idi Amin in 1971 should not only be viewed as a power struggle between the political and the military elites, but access to resources by one of the other of these groups. The second coming of Obote to power in 1979 after the Amin regime was routed out by the Tanzanian troops clearly demonstrated the irony of the conflict. Dictator Idi Amin had not only amassed the wealth of Uganda by expelling Asians and mismanaged resources, but initiated terror, human suffering and brain–drain and a total collapse of economic activity.

Thus between the period between 1966 and 1986 Uganda experienced a severe political and ecological stress as shown below:

1966 Expulsion of the Kabaka by Obote and initiation of regional conflict between the Northerners and Southerners.

1971 Expulsion of Obote by Idi Amin and initiation of social insecurity and ecological stress.

1979 Second coming of Obote to power and emergence of rival warlords for control of power and resources.

1985 Overthrow of Obote by Tito Okello and intensification of political conflict and ecological stress.

1986 Overthrow of Okello by Yoweri Museveni and emergence of religious war fanatics.

So, it can be seen that during those 20 years Uganda went through severe political turmoil. Social insecurity undermined law and order and led to mismanagement of life–support systems.

> "...political violence and economic mismanagement under Amin took toll of primary products. Coffee fell from 177,000 tons in 1970–1971 to 75,000 tons in 1978/79" (Christopher Stephens 1985).

Since agriculture is the main economic activity in Uganda the political conflicts could not allow farm activities to be performed in areas where guerilla activities were going on. It is unfortunate for Uganda that the conflicts were resolved at the cost of destruction of vegetation, wildlife, livestock, crops and water resources and human life (Mwanakatwe 1985).

During the conflict, for example, the flows of food and raw materials to the urban centres were either hampered or completely cut off. Similarly, because manufacturing virtually stopped, the rural poor were reduced to depend on the natural resources for their immediate needs. So many rural farms continued to revert to bush while the majority of the population both in the rural and urban areas slowly exhausted their food reserves. Fear reigned and frustration mounted. When Idi Amin at last fell in 1979, Uganda suffered the worst misappropriation of its public and privately owned resources through looting, hoarding, smuggling, etc. Moreover, the coming again of Obote to power in 1979 not only brought more misery but excerbated the situation especially from the guerilla activities! Freedom of access to and control of individual resources by individual persons ceased. This insecurity further triggered off massive migrations of Ugandans across the borders of Sudan, Zaire, Kenya and Rwanda.

The United Nations High Commissioner for Refugees estimated in 1985 that some 250,000 Ugandans did cross into the Sudan and another 4,000 into Kenya (UNHCR, 1986).

Since political boundaries are arbitrary and sometimes cut across the same ethnic groups, the influx of refugees precipitated further problems across the borders.

It was during this time that the people of the Karamajong and later of West Nile perished because of malnutrition, diseases, hunger and starvation. In Karamoja District alone, the increased drought and political insecurity caused severe famine which could only be contained by the International Charity Organizations.

> "Unlike parts of Northern Uganda where people depended on pastoral economy, southern Uganda's population was never threatened by the ordeal of hunger and starvation which was experienced in places such as Karamoja and later West Nile" (Alistair Matheson, 1982).

It should be pointed out that the Relief Aid had some negative and positive impact on the refugees. In Uganda, for example, returning refugees from Zaire Refugee Camps could not easily readjust themselves to their own environment after 5–10 years abandonment of their homes. Especially negative was the fact that many had to resort to indiscriminative use of the natural resources when the government had no immediate plans for resettling them during the transition period.

Environmental degradation and food insecurity during and after the wars

The economic development of a country usually relies on the health of the physical environment of that country. So when resources of a country are unsustainably and indiscriminately used, and the levels of exploitation become unacceptable, it can ultimately lead to the impairment of the ecosystem-functions and reduce the future capacity of the resources for human welfare.

Before establishing a relationship between the wars which took place in Uganda and the environmental damage they caused, it is important, first, to look at the nature of Uganda's physical environment. It is only then that one can attempt to establish a linkage between the two processes.

Land

There is no doubt that Uganda has a healthy environment. For this reason, the country has so far not appeared on the Hunger-List of the United Nations' Food and Agricultural Organization (FAO).

Even during the twenty years of political instability, Uganda's imports of food were never large enough to cause alarm, except perhaps for Karamoja which was rescued by the International Charity Organizations in the early 1980's.

About 22% of the country is estimated to possess very fertile soils. 43% is classified under "medium fertility" and only 2% is regarded as unproductive (Mwaka, 1988).

The civil wars which ravaged Uganda where mainly power struggles which, indirectly or directly, meant control of raw materials, food, energy and labour. It is very unfortunate for Uganda that the conflicts were resolved at the cost of so much destruction of the resources. It is doubtful if all the damages can be repaired. In areas like South Western Uganda, Acholi, West Nile, North Buganda (Lwero) and Soroti, to mention only a few, many forest and woodland areas were either shelled to destroy the enemy or scorched altogether by fire. Soils were also destroyed by bombs and grenades. Some rivers and lakes were either silted by physical earth transfers or soil sedimentation. The National Parks were exhausted of their wildlife and reverted to bush and were later recolonized by tsetse fly.

There are no figures to confirm the damage done by the 20-year civil wars, but it will definitely have some far-reaching impact on the environment of Uganda.

Documentary evidence already points out that Uganda, which was self-reliant in food, is beginning to face food problems because of mismanagement of the environment during the period of political instability.

Vegetation

Until the 1960s, about 47% of Uganda's landscape was covered with scattered trees and natural and cultural forests and woodland. Most of the latter was owned by the state.

From the estimated possible 10 million hectares of forest and woodland in 1980, only 1.6 million hectares was covered by the end of the civil war (Aluma, 1987). This figure represents only 6.3% of the total land area of 336,000 km, of which 194,000 km is land, 18% (42,000 km) open water and 4% swamp.

The most affected areas are, first, those where the wars were fought and second, those which were owned and guarded by the government. So catchment areas such as the Mabira Forest between Kampala and Jinja, the Budongo Forest in Masindi/Bunyoro, the Rwenzori Highlands, Kibale, and Maramagambo have lost their function as protective and productive areas.

Deterioration and deforestation are not new phenomena in Uganda. They took place also through the expansion of agricultural and settlement areas where population reached or passed the carrying capacity. Overgrazing was also another factor; so was the demand for firewood and building materials. But political instability accelerated these processes. In West Nile alone, almost all eucalyptus forest which were planted by the local government to cure tobacco and provide firewood have disappeared altogether through stealing and cultivation. It has been estimated that at the present rate of deforestation, Uganda will have no pure forest stands by the end of this century. The environmental impact of deforestation is that the relationship between vegetation and soil has been upset, i.e. the productive and protective function of the vegetation has been drastically reduced. In the same way, rainfall and radiation have been reduced and increased respectively. This will definitely affect the hydrological balance and initiate changes in underground water reserves.

Forests have been reservoirs of knowledge in Uganda. They have provided additional protein in the form of wild animals, rodents, reptiles, snails and birds. Their biochemical mechanisms have probably not yet been fully realized in the fields of medicine, food, fibres, fats, drugs, etc. It is doubtful if Uganda will ever recover from the losses brought about by the destruction of her vegetation.

Climate and water resources

Uganda has a favourable climate, a climate which has favoured a variety of food and cash crop production. It is perhaps this favourable climate which has all along made the country self-reliant in food. There are so far no meteorological assessments of climatic anomalies during the long period of war. However, local studies have revealed an overall random rainfall pattern over large areas (Mwaka, 1986). The changes in rainfall amount, distribution and variability in time and space have a direct impact on agriculture. For example, the Central West Nile District has in 1988 received very little, short-lived rain which came too late. As a result, the returning refugees—an estimated 20,000 people (UNHCR, 1986)—who have just returned and cultivated the land might face famine. Possible causes in decline in rainfall amount and variability are being related to the disappearance of vegetation and general global climatic anomalies. Temperatures in Kampala for instance have been perceived by the general public to be going up. Room temperature differences of 2°C and +3°C between 1970 and 1988 have been recorded in Kampala. Also Kabale and Mbale Highlands have recorded increases of 5°C between 1965 and 1988 (Mwaka, 1988). Although the changes reflected by the figures appear insignificant they can have far reaching effects on agriculture and therefore on food security.

As already stated above, 18% of the total area of the Uganda land-scape is open water and 4% is occupied by swamps. These wetlands occupy special importance in the general economic development of Uganda. The wars had a considerable impact on the wetlands. When the production system broke down, many industries closed down and the flow of imports of raw materials was hampered. This mostly affected the construction and building sector such as cement, iron sheet and steel. The absence of these materials forced the people of Uganda to resort to what nature offered. Wetland swamps in particular were turned into brick and tilemaking factories using clay. A journey from Kampala to Mbale near the Kenyan border would reveal that nearly every swamp the road passes has stocks of raw and burnt bricks. Brick-making, plus unplanned reclamation for vegetable production and pasture had virtually destroyed the wetlands of Uganda and reduced them to dry-land (Mwaka, 1988). Lake Kyoga—the largest wetland in the country—is now a dry lake. The environmental effects can be far reaching because wetlands play an important role in regulating water balance. They are homes for many forms of aquatic life and provide fish for additional protein and water for domestic and industrial use. Their disappearance can rob the country of extra agricultural land, scenic beauty and tourist attraction. The above activities are probably affecting all the wetlands in the country.

Population

Uganda's population has been increasing rapidly since Independence in 1962. There were nearly 7 million people at the time of independence. By 1969 the figure had gone up to 9 million and approached 12 million in 1980 and 14 million in 1982. Today, the total population has been estimated at 15.5 and projected at 25 million for the year 2000 AD (World Bank, 1984).

There is no doubt that these figures would have been much higher if civil war had not broken out and in fact it is surprising that the figure has been rising at all.

Human resources play a very crucial role in development and especially in changing the environment to benefit mankind through skill, labour and technology. The quality of a country's population is therefore very important for economic development.

Up to the 1960's, Uganda had a healthy population. However, after independence when political parties—U.P.C., DP and Kabaka Yekka (KY)—started resolving their political differences, Uganda began losing its human resources. For instance, the storming of the Kabaka's palace in 1966 forced many Kabaka government officials to flee with their Kabaka into exile. It is also not known how many people died or escaped during the overthrow of Obote's and Amin's regimes. The figure given by Amnesty International (1980), 500,000 people, may be an underestimate. Obote's second coming is estimated to have caused more human losses than Amin's regime. The massacres in the Lowero-Triangle, in Lango and West Nile are only a few of the discovered places of mass killings. It is also not exactly known how many escaped or were crippled by the wars.

The drain on human resources in Uganda is partly responsible for the collapse of the economy and deterioration of resources. It is estimated that some 5,000 trained Ugandans are selling their skills, labour and intelligence elsewhere in the World. This is a big investment. Unless persuasive efforts in the way of improving the infrastructure are made, Uganda will continue to suffer from a brain drain.

Already before the war, areas such as West Nile and Karamoja were experiencing environmental stress. Karamojongs had in the early 1980s overstretched the carrying capacity of their semi-arid land and caused extensive degradation through overgrazing. Today, this part of Uganda is seeing the beginning of desertification. Disease, hunger and malnutrition took their toll until international organizations came in to arrest the situation amid the political troubles. So the deterioration of the Ugandan environment is partly due to loss of human resources which would otherwise have developed the natural or life-support systems in Uganda and improve the general economic development.

MOZAMBIQUE: STARVATION BY DESIGN?

Mozambique attained its independence from Portugal in 1975 but this was after a protracted war of liberation that had been going on since 1964. Few countries have attained independence with such an appalling legacy of colonial neglect as Mozambique.

Since Portugal itself was the "sickman of Europe" at the time, with a fascist dictatorship in power, the situation in its colonial empire could only be deplorable.

At the time of independence, 93% of the Mozambican population was illiterate and more than 90% of its 10 million people lived out of reach of any form of health care or good water scheme. Of the more than 250,000 Portuguese and other expatriates who lived in Mozambique prior to independence, engaged in import and export trads and maintaining technical facilities, only 15,000 remained after independence.

But even with that situation in hand, the Mozambican Government did not want to pursue a selfish inward-looking policy. Bordering white-ruled Rhodesia, and conscious of its responsibility to the world as a frontline state, it imposed sanctions against Smith's Rhodesia, closed its borders, and allowed the Zimbabwean liberation movement to operate from its territory.

This is did in response to the resolutions of the OAU, the Non-Aligned Movement, and the United Nations Organisation of which it became a member after independence. The result of imposing sanctions against Rhodesia was that it denied itself payments which it would have collected had the landlocked Rhodesian regime used the Mozambican routes to reach the sea; not only that, but it also invited military incursions into its territory from the Rhodesian rebel forces.

At the time the international community manifested only verbal sympathy towards Mozambique. Britain, which was the colonial power in Rhodesia and whose responsibility it was to crush the Rhodesian rebellion, did nothing to help Mozambique, neither did a number of Western countries that claimed to be the defenders of freedom and peace.

One had hoped that with the independence of Zimbabwe peace would come to Mozambique, but hardly a week has passed in peace. The armed bandits that were created by the Rhodesian intelligence system to fight against the Zimbabwe's liberation movements were sent, after Zimbabwe's independence, for retraining in South Africa to be used against independent Mozambique.

The MNR in Mozambique is supplied, equipped, trained, paid, transported, directed, and on occasion led, by the Pretoria regime. The present famine in Mozambique is therefore the direct result of the destablisation policy of the Pretoria regime which used mass terrorism through its proxy organisation—the MNR.

Mozambique has recently estimated that about 100,000 people have died as a result of the activities of MNR. This was not by bad weather or a misdirected agricultural policy, but rather by the MNR's disruption of rural life and food production in Mozambique, its destruction of Mozambique's ability to raise export earnings, to purchase food and distribute it, and its interference in or outright prevention of food relief distribution.

The effect of the present war of destabilisation in Mozambique is that 45% of the deaths of under-fives are attributable to the impact of the war. The war's victims are not only economic and military targets but also the very social fabric of the nation.

The deaths stem from the destruction of health and education facilities, dislocation of communities, disruption of agricultural activities and the construction of health and water budgets as a result of the war.

The share of Mozambique's national budget that goes to recurrent defence costs is 42%, one of the highest in the world; and because it is forced to spend so much on the war effort it is unable to meet the costs of other services.

As a result of this war of destabilisation being conducted by the Pretoria regime against Mozambique, the country now has one-third of its population facing starvation; 10% of its people are homeless; half a million are refugees in their own country; half of the schools are either closed or destroyed, as are one-third of the clinics and health centres; and this country at war is 3.4 billion US dollars in debt.

The question is: why is South Africa so persistent in destabilising not only Mozambique but also Angola and other countries of the Southern African region?

The fall of the Portuguese empire and the independence of Zimbabwe together with the support of the progressive forces in the international community gave a strong and new impetus to the South African liberation movement by removing the buffer zone of South Africa.

Now South Africa and Namibia share a common border with independent African states. The frontline has now moved further south— the frontiers of the liberation struggle are the Namibian border with Angola, and the South African borders with Botswana, Mozambique and Zimbabwe, and the frontline states now include Tanzania, Botswana, Zambia and Zimbabwe.

Because of this shift of the battle-lines, the South African regime panicked and intensified its internal and external struggle for survival. The Afrikaaner establishment was forced to choose whether to adopt the traditional 'laager' concept or to take the offensive posture by pushing its area of influence northwards as far as possible and taking advantage of its economic power in the region.

The worsening economic situation in the region encouraged the Pretoria regime to embark on a policy of destabilisation which is both economic and military. South Africa unleashed attacks on Angola, Mozambique and the other countries in the Southern African region.

Due to an identity of interest with the South Africa regime, the major Western countries are in no position to assist the Southern African states to resist military and economic attacks. Solidarity rhetoric and slogans cannot be a substitute for effective and coordinated economic and military support.

Thus South Africa's ruthless acts of destabilisation and aggression against Mozambique unfortunately led Mozambique to the Nkomati Accord. Pretoria considered such short-term success as a diplomatic triumph, but that did not stop its international isolation.

It is true that agreements with Mozambique and Swaziland affected the capability of the ANC in channeling resources and cadres into the apartheid regime but that has not affected the struggle inside the Boer Republic where the opposition against apartheid has become better organised, more widespread and more militant.

Everyone now realises that the root cause of the problem in Southern Africa is the apartheid system within South Africa.

We are now living in a world of interdependence: whatever happens in one part of the world affects other parts of the world. The Chernobyl incident has shown that the concern on the effect of nuclear weapons goes beyond the nuclear power themselves.

Finland and Sweden, Poland and Holland, could see the radioactive elements spilling into their territories all the way from the Ukraine. During the American war of aggression against Vietnam, American mothers and wives were in the same situation as those in Vietnam when they started receiving the coffins of their beloved ones.

International events in the last few years have proved once more the point that there can be no nation which can accept humiliation and oppression, and that in fact the nation which oppresses another nation forgets its own chains.

Southern Africa has been plunged into a war situation because of the fact that South Africa seeks to make that region safe for apartheid and profitable to the apartheid economy, but Britain and its Western allies cannot claim that they are not parties to this war.

British investments in South Africa far exceed her total investments in the rest of Africa. The extraction of resources from South Africa by British companies is a well-known fact. Against the United Nations recommendation, the governments of West Germany, Holland and Britain extract uranium from Namibia despite the fact that Namibia is illegally occupied by South Africa.

Western powers and Israel have offered all-out military support to the South African regime and have enabled Pretoria to develop its military and nuclear industry. Because of the United States so-called constructive engagement policy and the acquiescence of the British Government to that policy, the racist South African regime opposes decisions of the international community on Namibia.

The war which South Africa has imposed on the countries of Southern Africa has dislocated the economies of these countries and displaced their population. The economic cost of this destruction in these countries is estimated to be over 28 billion US dollars.

Those are the reasons why Mozambique today is facing famine. Official statistics point out that about 4 million people cannot be fed from domestic production and are in grave danger of extreme malnutrition or starvation in the absence of very high levels of food aid.

But people are dying in Mozambique by design, by the plot of the South African regime with the acquiescence of its Western backers.

CONCLUSION

The above two cases have illustrated in detail how political conflicts can lead to ecological stress. Whether we are establishing an obvious relationship between political conflict and ecological stress, we must bear in mind that the two processes are two sides of the same coin. The on-going events in Uganda and Mozambique have definitely sown the seed for deterioration of resources which eventually lead to ecological stress and aggravate further conflict. Whereas the wars in Uganda had the nature of tribal, religious and power struggle, those in Mozambique have been engineered and designed by the racist South African regime.

The overall environmental and economic consequences in the two countries are almost identical—social destabilisation, loss of lives and property, mismanagement of resources which has led to collapse of economic activities and food insecurity. While in Uganda the UNHCR and some government and charity organizations tried to resettle people across borders permanently or temporarily, this has not been the case in Mozambique. In the absence of any help forthcoming, political insecurity and ecological stress tend to intensify each other. This is what is happening in Southern Africa.

Nevertheless, caution should be taken by those bodies like UNHCR which try to help through relief aid, for these displaced people may fail to readjust and cater for themselves after the conflicts.

In Uganda the conflicts have left deep imprints especially on the very young who have not only witnesses the sufferings but took part in the conflict.

In Southern Africa, and in Mozambique in particular, the apartheid policy of destabilization is slowly manifesting itself in areas which have more serious political and environmental consequences on the Frontline States.

The whole situation in Africa and in Southern Africa in particular therefore calls for an international cooperation to resolve the human rights issue so that a degree of access to and control of resources among individuals and governments is achieved and maintained.

REFERENCES

Aluma, T.R.W. 1987: *Deforestation and Agricultural Drought with reference to Uganda*, Workshop on Drought mitigation, Kampala.

Coetzee, D. 1977: *Is the Sahel Drought Coming back again?* New African Development Magazine, England, p. 81

Matheson, Alistair, 1982: *Uganda*. African Guide. World Information Service. 6th Year of Publication.

Mwaka, V.M. 1988: *The Environment and Food Security in Uganda*. Paper prepared for the IUC Geography Workshop at Mikumi, Tanzania.

Mwanakatwe, K. 1985: *Uganda Coup after coup*. M.A. Thesis University of Dar es Salaam (unpublished).

Othman, H. and Kiss, J. 1986: The Destabilising Role of the Apartheid Regime, Enfoques No. 12, 1986, Havana.

Othman, H. 1988: *Mozambique. Starvation by Design?* In: Sunday News, January, 10th, 1988.

Our Common Future, 1987: The World Commission on Environment and Development. Oxford University Press.

Stephens, Christopher, 1981: *Uganda..* African Guide. World Information Magazine, England. pp. 381–386 and pp. 405–408.

Timberlake, L. 1987: *Krisenkontinent Afrika. Der Umwelt Bankrott. Ursachen und Abwendung.* Peter Hammer Verlag Ole, Bonn.

UNHCR, 1986: Refugee: *Ugandans in Sudan and Kenya*. (Dossier) No. 29.

UNHCR, 1987: *Refugees and Displaced Persons from Mozambique*. (Dossier) No. 37.

World Bank Report, 1984, (Africa).

The Rwandese Refugees in Uganda

Byaruga Emansueto Foster

INTRODUCTION

The aim of this paper is to demonstrate that the Rwandese refugees have had a profound impact on the Uganda environment and have affected its socio-economic conditions. The impact is prevalent in the Western parts of Uganda in particular where they have settled. These are *Kyangwali* in Bunyoro, *Rwarwanja, Ibuga (Kasese), Kyvaka* I and II and *Kahunge* in Tooro and lastly *Oruchinga* and *Nakivale* in Ankole. To a lesser extent, their impact is felt in other areas of Uganda in general. The arrival of the Rwandese and their activities within the country, the activities of the indigenous people who inhabited the areas (the invaded), the activities of the Rwandese who remained in Rwanda, the reaction of the governments of Uganda and Rwanda as well as the activities of the international community will be discussed in order to understand their consequences on the environment and their contribution to political conflicts in Uganda.

Environmental stress in this paper will refer to disturbances on the surroundings of a person. This can vary from air, water, rivers, flowers, grass, forests insects, animals, game reserves and public parks. It also includes stress among human beings. Clearly, environmental stress cuts across the board, ranging from physical, material, social to economic and political aspects. The Rwandese refugees have invaded all these aspects in Western Uganda, which invasion has resulted into political aspects. The gravity of the conflict has varied, overtimes depending on the changing relationships between the refugees and the nationals, and the political situation of the country as a whole. We shall use a historical analysis in the exercise and at the end of the paper, we will suggest solutions.

BACKGROUND TO THE REFUGEE PROBLEM IN RWANDA

Rwanda is a small country in the south-western part of Uganda. In 1986 it had a population of 6.27 million on 238 square km. The population density is the highest in mainland Africa, yet 10% of Rwanda is dedicated to parks. It has a population growth of 3.2% annually. The population is 85% Bahutu, 14% Batutsi and 1% Batwa (pygmy type).

Rwanda has fertile soils, rich forests and a very good climate. But due to population pressure, land is very scarce and causes a lot of conflicts between the cattle, 82,000 pigs, 338,000 sheep and 1, 157,000 goats. The high population together with a high population of livestock make Rwanda a very small country for one to live in comfortably.

Before colonialism, the political system was feudo-monarchical and the rulers came from the Tutsi minority ethnic group. Some few Bahutu rich farmers succeeded in getting favours from the ruling Tutsi class but they did not become rulers although they lived under less harsh conditions. The point being made is that the rulers were the Batutsi, and the Bahutu were serfs. When the colonizers arrived, they preferred to rely on the existing socio-political structures and used indirect rule by using the dominating minority. The Batutsi were therefore dispensed from forced labour and only acted as implementors of colonial policy. During the fifties, the elites of the dominated Bahutu started to agitate for equal opportunities in jobs, education and the running of the country. They formed an association to articulate their interests and on 24 March 1957 they published the *Bahutu Manifesto*. A year later, on 30 March 1958, the King appointed 10 Bahutu and 10 Batutsi to study the nature of the social conflict between themselves. In mid-April 1958, a meeting of 40 chiefs, all Tutsi, met to discuss the attitude of the Bahutu elites whom they criticized strongly. In June 1958 the report of the Committee set up by the King was discussed by the King's Supreme Council where six Bahutu were invited as observers. Most of the discussion was spent by the King who castigated and condemned Bahutu for disloyalty and ingratitude. The Council concluded that there were no problems between Bahutu and Batutsi in Rwanda and that those who brought up the problem were malicious self-seekers who deserved public condemnation as criminals.

King Rudahigwa died on 25 July 1959 before he solved the problem that exasperated over time each camp; where one tried to keep the priviliges of ruling while the other tried to fight the injustice to acquire equitable representation and fair distribution of natural resources. The new king, *Kigeli*, became heir to the throne and declared that he wanted to be a constitutional king. This decision ushered the formation of political parties. Three parties appeared on the scene: UNAR—a monarchist party which wanted to maintain the status quo, RADER, and APROSOMA which were somehow confused with regard to their real objectives. In October 1959 PARMEHUTU was declared. It was a Bahutu party which recognized the constitutional king as long as he supported democracy and progressive ideals of the party. UNAR (Union National Rwandaise) started a campaign of intimidations towards other parties accusing them (especially PARMEHUTU) of being anti-King. It demanded national independence immediately from the Belgians. It campaigned among the so-called "progressive" countries at the United Nations, championing to be the real national party, accusing the others to be neocolonialists. Among its membership were all the chiefs and government officials. Hence all government machinery and logistics supported the party. On the other hand, the PARMEHUTU was saying "Democracy first and independence after". The conflict became great, especially during the month of October 1959, between the Batutsi and the Bahutu.

The fire was sparked off on 1 November 1959 when four young Batutsi attacked *Mbonyumutwa Dominique*, who was a Bahutu leader, on his way from church service. The cry of his wife for help was interpreted that he had been killed. The enraged Bahutu started setting fire. Some people ran to other areas which had not experienced the same troubles and some went outside Rwanda. The Batutsi also fought back. They started murdering Bahutu leaders in *Nyanza*, *Gitarama* and *Astrida*. On 6 November 1959 the King Kigeli sent a telegram to King *Baudouin* proposing that since the Belgians had failed to restore order, he should be allowed to use his militia. Before receiving a reply, his militia was deployed against the Bahutu on the side of UNAR. The Belgian army stepped in and stopped the civil war on 12 November 1959. 1,240 people were arrested for involvement in the troubles. 62% were sentenced between 1–5 years jail terms, 7% between 5–10 years, 3% for more than 10 years and two people were sentenced to death regardless of their ethnic background.

On 10 November 1959, Belgium declared that it wanted political changes that would lead to self-government in Rwanda although the changes still recognized the King, they advocated for elections of Gombolola (sub-county) chiefs and counsellors. UNAR decided to boycott these elections which took place on 26 June to 31 July 1961. After the elections, PARMEHUTU had won 70.4% of the seats. It is these elections that led to Rwanda's self independence internationally on 26 October 1960. On the 28 Jan. 1961, the Minister for International Affairs called a meeting of all Gombolola chiefs and counsellors. They were 3,126 in number, of which 229 were chiefs. They met at Gitrama where they unanimously voted to abolish monarchy and established a republic. This came to be known as the coup d'état of Gitarama which put an end to the activities of UNAR. On 25 September 1961, a referendum was organized under the auspices of the United Nations to choose between monarchy and republic. 80% voted for a republic and King Kigeli decided immediately to leave the country together with his sympathizers. Some went to Burundi, Congo (Zaire) and Uganda. The number of refugees in Uganda today is estimated to be 84,000 people. It is this invasion of the Rwandese refugees that has caused public concern, especially among the local people in the invaded areas, the Uganda administration and the United Nations.

THE RWANDESE REFUGEES IN UGANDA

Before the coup of 1961, Uganda did not have Rwandese as refugees. They were immigrant labourers who were free to move into Uganda in search for manual work. But with the influx of the refugees after the upheavals of 1961, four issues became important. Firstly, there was a need to control the Uganda–Rwanda border in order to keep trace of these refugees. Hitherto, there had been no need to police the borders. Borders had been looked at as superimposed colonial structures that divided mutually related people. But now border immigration control became an important issue.

The control of the borders did not stop the influx. At a speech to the Council of Ministers, the Minister of Security and External Relations said:

> "...If, as anticipated, the Bahutu were victorious in the elections, between ten and twenty thousand more refugees were to cross the Rwanda border into Uganda... some 3,000 refugees (largely Batutsi) were already in Kigezi after the 1960 exodus... some 400 refugees a day were already trying to enter Uganda..."[1]

Hence the arrival of the Rwandese in such large numbers created problems of immigration control. To allow them to enter, government had to create refugee settlements, hire personnel and appoint a Director of Refugees who would in turn create reception camps to deal with issues related to accomodation, resettlement and security. But this expansion in services created and increased strains on the administrative and technical staff, let alone the local population. Areas mainly affected were agriculture, veterinary, health, local administration, social service and welfare. These had to pull resources together in order to deal with the influx. For these areas to deal effectively with the refugee problems, they needed money. The money was not available. They needed food which was also not available. They required land, reception facilities, transport, food and food distribution logistics and new medical centres. Problems hitherto unknown in the western part of Uganda became a reality. These included camps, toilets, garbage collection, dispensaries, veterinary clinics, storage facilities and roads. Government had to appeal to external help and the United Nations High

---

[1] Jacob, B.L. *Administrators in East Africa*. The Government Printer, Entebbe, Uganda, 1965, pp. 9-10

Commissioner for refugees (UNHCR) came in. With the help of UNHCR, the following centres (camps) are now operational:

Kyangwali	In *Bunyro*
Rwarwanja	In *Tooro*
Ibuga (Kasese)	"-
Kyaka I	"-
Kyaka II (population 17,621)	"-
Kachunge	"-
Oruchinga	In *Ankole*
Nakivale	"-

UNHCR is silent on the population in each camp but the total is known to be 84,000.

The second issue regarded cattle grazing and the availability of grass. The Rwandese are cattle-keeping people. They seasonly migrated and wandered in search of water, grass and even to avoid tsetse-fly infected areas. This means that they moved into and out of Uganda depending on the needs of their cattle. Since this movement was historical, it caused less pressure and tension on the people and the environment. But now the Rwandese were flowing into the country with their cattle to overpopulate an already overpopulated cattle area. As all cattle keepers behave, they do not want to de-herd. The government asked them to sell off some of their beasts. They refused resulting in conflict between the government and themselves. Because of the large numbers of cattle coming into the country unexpectedly, the following happened:

Firstly there was need to encroach on the traditionally non-cattle keeping areas. But some of these areas had tsetse-flies which had to be cleared. But due to pressure, cattle moved in before they were cleared. This brought in the problems of trypanosomiasis which in turn demanded veterinary attention. All these required money to be cleared. The government hoped that this would enable it to keep law and order, be vigilant in assessing who was a refugee and who was not and force them to sell off the excess cattle. The consequences of this decision are obvious in terms of environmental strain. The Rwandese refused to sell off the cattle, resulting in overherding, overgrazing and spreading of cattle diseases like trypanosomiasis, blackwater, anthrax, foot and mouth disease, murderpest and bovine pleuro pneumonia. The Rwandese cattle were also bringing diseases to areas which did not have them before. The Director of Veterinary Services lamented:

"... There was a very live danger of bovine pleuro pneumonia or murderpest being brought to areas which had been free of such diseases for years... " The veterinary services in Rwanda no longer existed. One third of Ankole was infected with tse tse flies which had advanced very recently requiring 45,000 head of cattle to be vaccinated.[1]

Thousands of cattle died due to diseases and lack of pasture. Also, in the non-cattle-keeping areas, the animals brought in cattle flies. The indigenous people were not used to such flies and hated them. Also the cattle often grazed on the crops of the invaded. Physical confrontation became a necessity and deaths resulting from such action were reported.

Secondly, in the cattle-keeping areas, apart from the coming in of large herds, there was competition for grass and water. Grazing land became a major problem as the cattle competed for grass. There was also competition for land to grow food between the invaders and the invaded. Consequently, fights became regular, the topography was trampled and the vegetation cover changed. Forests were invaded and the game parks were invaded too. This had not been the case before the refugees came in.

The third issue regarding the Rwandese refugees concerns social matters. Historically, a part of Kigezi district called Bufumbira in Kigezi was alright. They could go there and stay until conditions improved at home.

The fourth issue, in Rwandese royal traditions, the king of Rwanda married from the royal family of the Ankole kings or Tooro royal family. In fact, the present heir to the throne of Tooro is married to a beautiful princess from Rwanda. The Batutsi Rwandese were therefore cousins of the Ankole or Tooro royal families. In the African traditional society context therefore, the Batutsi and the Banyankole (Bahima) royal families were so close that each was ready to help the other in case of need. Two important points must be noted here. The first is concerned with the relationship between the Ankole people and the Batutsi. In Ankole there are two ethnic groups: The Bahima and the Bairu. The Bahima were the traditional rulers while the Bairus were the serfs, like the Bahutu in Rwanda. Traditionally, though now disappearing, there have been conflicts between the ruling Bahima and the ruled Bairu. So, whereas the Bahima were willing to allow the Batutsi to come in, the Bairu saw them as invaders who had to be fought and thrown out. The Batutsi were coming in to join hands with the Bahima to take away the little land belonging to Bairu. Also the coming of the Batutsi with their cattle was to bring in new cattle diseases to kill local cattle: Hence the more grounds for them to go

---

[1] *Ibid*, p 45.

away. The second point concerns the "nearness" of these Batutsi to the borders of Rwanda. Earlier on we indicated that the Batutsi supported NAR. Even when they ran away, they did not drop these objectives. Therefore it is not in the government of Rwanda's interest to have them near the border. In fact, at the beginning of December 1963, a surprise attack was organized from Burundi and Uganda to get supplies from border gombololas. Gombolola chiefs and counsellors were killed and the attack stopped only 20 km from the capital, having overrun the small barracks at Gako. The Uganda government realized this need and requested them to move to Nakiwali further away from the border. But the Rwandese did not want to move until they were forced to do so. The reasons they advanced were that they felt more at home in the areas surrounding the Rwanda border. Secondly, they felt at home among the Bahima and could easily get news from Rwanda and thereby be able to influence political events in Rwanda. Lastly, although this area was fertile, it was not fit for animals because it was infected with tsetse-flies which would cause trypanosomiasis to animals and sleeping sickness among people. Despite this reasoning, the government took the expensive decision to remove the tsetse through sprays and cutting the grass, providing veterinary and medical facilities in the area and asking the international community for help. They settled there and some participated actively in the politics that shaped present day Uganda.

THE ACTIVITIES OF UNHCR AND INTERNATIONAL AGENCIES

Facilities to get information about the UNHCR activities in the seven areas of Rwanda refugees were futile. But the following information filtered through regarding refugees in Kasese and Kabarole. It was learnt that the Uganda government together with the UNHCR want all refugees to be self-reliant. To be self-reliant, certain infrastructural arrangements have to be made. The UNHCR is very willing to help but it is being let down by the successive Ugandan governments who have tended to say rather than to do. Because of this, all UNHCR efforts have been frustrated. In Kyaka, I, for example, a respondent remembers with nostalgia what he used to get from UNHCR in 1964. In addition to food, their cattle were taken care of, schools were built plus houses for the settlement staff. UNHCR also supplied them with a tractor to facilitate opening the land. Health services were supplemented with an ambulance. Transport was done by a settlement lorry and the senior staff had a landrover. But the UNHCR could not take care of the maintenance expenses. They were left to the Uganda government. Today, a fully fledged primary school still exists, but in bad shape. A dispensary is falling to pieces. Staff houses badly need repair. The tractor broke down about two years ago. The ambulance is non-existent and there is no lorry. The road to Kyegegwa from Rityana is unpassable. From Kyegegwa to the camp it is safer to walk than try to drive. Veterinary services have disappeared and the future of the cattle is in balance. Tsetse flies are back. Tsetse control picket lines are seen every 10 kilometers but there is no drug. When it is available, it is too expensive. The explanation given for this development is that Obote governments were always against the Rwandese refugees since the 1960s. The respondent remembers vividly when Mr Katiiti CB used to visit them and abuse them publicly. During the second Obote regime relations were so hostile that they culminated into overt persecution of Rwandese refugees accusing them, inter alia, of having supported Amin.

The example of Kyaka gives us what we expect to see in other camps: A state of deterioration due to gross mismanagement. But we also understood that the UNHCR had to be cautious in dealing with the refugees because of its very nature of functioning.

We cannot blame the UNHCR especially when we examine what happened in 1983. During Obote II period, the then Minister of International Affairs, John Luwuliza Kirundo, introduced a bill in the parliament entitled "The Alien Registration and Control Bill". The bill intended to empower the Minister to decide on who was a citizen, no matter what colour or creed. No person could decide by himself whether he was a Ugandan unless he had legally processed the documents to prove so. An alien was defined as a person who was not a Ugandan, a foreigner who was not a subject of the country where his ancestors were born. Hence a naturalized person became an alien and a non-citizen. All aliens were to be issued with cards within seven days of their registration at the District Offices and all immigration questions had to be resolved at the Attorney General's Chambers.

The motives behind this move were political. The Uganda's People's Congress had lost the 1980 elections although it had taken over government. Among the many reasons why it had lost, especially in the West, was that the Rwandese refugees were predominantly supporters of the Democratic Party. To allow them to continue participating in the politics of Uganda was to hang UPC as a party. Secondly, Yoweri Kaguta Museveni, the then leader of Uganda Patriotic Movement (UPM) had gone into the "bush" to wage a guerilla war against Milton Obote's government which had openly rigged the elections. Most of his supporters in the "bush" were believed to be the Rwandese Democratic Party supporters. In order to discredit Museveni and his struggle, the government had to tell the Ugandans that Museveni was a Rwandese and his supporters in the struggle were Banyarwanda and therefore non-Ugandans. It was non-Ugandans fighting Ugandans. The government thought that appealing to this sentiment would unite Ugandans to join the Obote forces to fight Museveni. The Ugandans, fortunately, knew better. Thirdly, the bill though openly denying it, tacitly encouraged the people in areas where the Rwandese refugees had settled to harass them. With one week from the date of the presentation of the bill, the natives had displaced the Rwandese refugees and grabbed their properties. Although the Rwandese refugees were scattered all over the country, the most affected were those in Rakai, Masaka and Mbarara districts where the UPC Youthwingers took it upon themselves to hunt and punish them.

The activities of the Youthwingers were nakedly brutal. They killed, raped and grabbed property with impunity. The UNHCR was accused of giving relief aid to rebels of the National Resistance Army. Hence it became impossible to bring drugs to the settlements. The activities of the Youthwingers put the Rwandese refugees who had settled in Uganda since the 1960s on the run for their lives. The problem however was where do they go now? They could not return to Rwanda. The Rwanda government had made it clear that it had no space for them. It claimed that since their departure, the population of Rwanda had quadruplicated. The Rwanda government also thought that their return would re-rupture a politically healing wound of the 1960s. They could not go to Burundi because there was internal strife there. They could not go to Zaire because Mobutu did not want them either. The only alternative was to join Museveni in the "bush" war. This became the only survival outlet and this they did effectively. But having stormed Kampala with Museveni's Resistance Army and succeeded in taking over the reigns of power, the problem is emerging again. The problem seems to be now in the army ranks especially in terms of who leads what and who commands what regiment.

But with the coming of a new regime that cares, the MSF has decided to salvage the camps. For example, Dr Fred Kruiswijk who is in charge of the medical services in Kyaka II is reported to have said that plans were under way to set up new permanent structures. At the moment they were concentrating on immunization and other forms of health care and education. The only problem he had was shortage of trained staff. He was embarking on training local staff with his medical team (Médecins Sans Frontières). The problem we see here is that when MSF pulls out, the Uganda government will not be able to employ the locally trained staff. It does not have the money. Also the refugees are now having the privilege of luxury drugs which no other Ugandan, except a visitor to the camp, can have. The refugee is now a privileged man until MSF pulls out.

The UNHCR is promoting the creation of cooperative societies for self-reliance of the refugees. To these societies, the UNHCR intends to provide masonry tool boxes, carpentry tool boxes, pre-pack brick making machines so that the refugees can build their own schools, dispensaries, staff houses, bore holes, valley tanks, regrade roads and make money to buy their own books. The cost is estimated to be US$ 1,665,598. The problem here is that the locals will begin to see the Rwandese refugees as the privileged foreigners on Ugandan soil. He hopes that this will not happen if the Ugandan government follows what we are suggesting below.

SUGGESTIONS

Presently, the Rwandese refugee problem is seen more in its political, administrative and social form than its historical context. We already know from the paper that the problem is older than the independence governments. But also know that societies cannot be observed and analyzed in a static way. It is true we call them Rwandese refugees because most of us were alive when they came. But how sure are we that we are the true Ugandans? We, according to history, came from somewhere else outside Uganda. Then who has the power to decide who is a Ugandan and who is not? We suggest that since these people have been with us for over twenty-five years and have settled, have inter-married locally and since they have participated effectively in the transformation of the country, they should be left and accepted as Ugandans. On their part, they should also stop meddling in the affairs of the Rwanda government. The hope of returning home by the Batutsi must be abandoned. In any case, the situation in Rwanda, its size, wealth and population dictate against the idea. Secondly, arising from the first proposal, it becomes important for the Uganda government to have a clear policy on immigration and laws, processes and structures which encourage absorption of the Rwandese community in Uganda. Thirdly, the Rwandese refugee problem cannot be discussed in terms of refugees but in terms of the general environmental dictates existing. If they are causing environmental stress in the Western parts of Uganda, they should move to the less populated areas of the country. For example, the Bakiga have moved to Bunyoro. The Bunyoro have occupied Bulemeezi and Ssingo countries. There is no reason why the Rwandese cannot move too, instead of inviting Maragolis from Kenya and the Palestinians. Fourthly, in order to reduce environmental stress and political conflicts, international organizations, working hand in hand with the governments of Uganda and Rwanda should educate the Rwandese to understand and appreciate the chang-

ing situations. This process of education, mobilizing and politicalization should include Ugandans too. We have two options. Either ac-
cepting the "Salad bowl" or the "Melting pot". The authors prefer the
latter rather than the former. We also believe that the Museveni Sec-
tarian Bill is a step towards this direction. It is a step towards the building of a nation based on the doctrine of gradual integration and
national unity. Lastly, it is very important for the UNHCR, especially
that in Uganda, to be at the forefront in this effort. Also it should deal
with the problem in a lesser secretive manner. Presently, it is a closed
source of information on the problem. It does not allow researchers to
examine its activities in Uganda. If these issues are rectified, we can see
a change in the attitude, a disappearance of "hauntism" among the
Rwandese–Ugandans and the withering away of environmental stress
and political conflicts caused by refugees of Rwanda origin in Uganda.

The Ishaq–Ogaden Dispute

John Markakis

"There is no peace with the Ogaden" according to an Ishaq saying that refers to an old and bitter conflict between these two pastoralist Somali clans. Naturally, the bone of contention is land, and the intensity of the conflict has kept pace with the continuous degradation of the terrain in the region; a malignant process promoted by economic and social as well as natural factors. Furthermore, the intensity and destructiveness of the internecine struggle was greatly exacerbated by the intervention of political forces from outside the pastoralist realm, and whereas earlier hostilities were spontaneous and occasional and claimed relatively few victims, nowadays a regular war is fought with modern weapons, threatening to depopulate the contested region. This article examines the fateful relationship between ecological stress and political conflict in a region of the Horn of Africa which comprises an economic unit fragmented by the Ethiopian–Somali border.

The Ogaden region is known as the Haud (Somali for 'south') and covers about 40,000 square kilometres. It lies inside the Ogaden alongside the boundary with northern Somalia. Called the 'waterless Haud', it has no permanent source of water, though it is crossed by two seasonal streams called Tug Jerer and Tug Fafan. A high undulating plateau with wide grass plains, the Haud's vegetation includes thorn bush and aloes and its plains are dotted with giant anthills. During the rainy season that comes twice a year, the Haud plains turn into seas of grass and provide the best and largest grazing area in that part of the Horn. At this time the Haud is visited by several Somali clans from both sides of the border. Feeding on green pasture, the herds of camels, sheep and goats need no water and the herders could stay up to three months each visit, before returning to their base areas and permanent water sources for the dry season. The absence of permanent water sources in the Haud is a blessing in disguise, because it rules out year round occupation and continuous grazing. This allowed its pastures to recover completely after each season and to preserve and regenerate their full potential.

The Haud lies inside the Ogaden, a vast sloping plain of about 200,000 square kilometres that extends from the southern reaches of the Harar plateau in the north to the Ethiopian Somali border in the east and south and the Webi Shebeli river in the west. The region takes its name, the Ogaden, from the dominant Somali clan that lives there. The Ogaden, like many of their kinsmen, are camel, sheep and goat herders. They reside entirely within the Ogaden and have no need to venture outside its borders to find pastures. The northeastern corner of the Ogaden, the Jijiga plain, where rainfed cultivation is possible, is inhabited by small clans, like the Bartire and the Gerri, many of whom practise agriculture.

Several other clans whose base areas are found outside the Ogaden visit it regularly for seasonal grazing, especially in the Haud. The Ishaq from the northeast, the Dolbahanda from the southeast, the Marehan and the Beidyhan from the south, share the prized pasture-land with the Ogaden. The congregation of these clans in one area naturally increased the incidence of mutual raising for animals, a traditional practice that led to occasional bloodshed and served to maintain a permanent state of tension between the different clans and clan families. This was the normal state of affairs in Somali pastoralist society, where a native poet complained that "Peace worsens the condition of my household".[1] Traditional institutions designed to contain and resolve this type of conflict functioned effectively. The most important of these was the *diya*. a contractual group which paid and received blood compensation for injury done and sustained by its members. In the early 1950s, the standard rate of compensation for the killing of an adult male was one hundred camels.[2]

The Ishaq, the dominant clan family in northern Somalia across the border from the Ogaden, have always been the most dependent on the Haud. Without any rivers and an annual average rainfall of less than ten inches, northern Somalia's resources are insufficient for the needs of a population that is roughly double the population of the Ogaden and for an animal population of a comparable size. Predominantly camel, sheep and goat herders, the Ishaq are also intrepid traders and caravaneers who always controlled trade routes between the Ogaden, Hargeisa and the port of Berbera; a privilege that did not endear them to the Ogaden who have had to buy safe passage along those routes. The Ogaden retaliated by raiding Ishaq herds and flocks in the Haud. further embittering relations between them. Since neither of them developed centralised political structures, hostilities between them remained episodic and low key. Centralised state structures encapsulated the Somali nomads during the last quarter of the nineteenth century. The imperialist scramble partitioned their lands among no less than five states. Four of these were European colonial creations belonging to Britain (Northern Somaliland, Northern Kenya). Italy (Southern Somalia). and France (Djibouti). The fifth was the expanding Abyssinian kingdom, henceforth to be known as the Ethiopian Empire which claimed the Ogaden and adjacent Somali lands west of the Webi Shebeli.

While the scramble was on, European and Ethiopian claims over-lapped extensively, and lengthy negotiations were required before the boundaries were demarkated. They roughly comprise today's boundary lines. Britain claimed a portion of the Ogaden, including the Haud, on the grounds that the Ogaden pastures were essential for its Somali subjects. Ethiopia claimed an analogous portion of British Somaliland on the grounds of prior conquest. Emperor Menelik's negotiating position was strong, for the Ethiopians had just defeated the Italians in 1896 at the battle of Adua. The British were involved in a war against the Madhist state in the Sudan, and were exceedingly apprehensive about the possibility of an Ethiopian–Sudanese alliance. Therefore, they chose compromise, and finally surrendered about one-third of the area of their Protectorate. The dividing line that placed the Haud within Ethiopia was drawn by Ras Makonnen, Haile Selassie's father the then Governor of Harar. The 1897 Anglo–Ethiopian Agreement included a clause safeguarding the rights of pastoralists, like the Ishaq, who found themselves under British rule, to use the Haud pastures across the border. This right was not interfered with for several decades. Ethiopia was still a feudal polity and showed little interest in the pastoralist low-lands where no easily exploitable source of wealth could be found. No attempt was made to police the Ogaden, and Ethiopian garrisons did not appear south of Jijiga until the 1930s, when the Italian menace appeared. Indeed, the border itself was not marked, and the nomads were scarcely aware that it existed. When an Anglo–Ethiopian Commission was demarkating the boundary in 1931, it was attacked by the Somali and one of its members was killed. The border was a serious obstacle to the pastoralists of the British colony because it impeded their entry to the Haud. The Ogaden, on the other hand, were increasingly resentful of such intrusions in what they had come to consider their own territory, and bemoaned Ethiopia's acquiescence to them. When Haile Selassie came to Harar in 1935 to rally the Somali against the impending Italian attack, he heard lengthy complaints from the Ogaden chiefs, including a denounciation of the agreement that allowed the nomads from across the border to enter the Ogaden. Makhtal Dahir, a youthful chief who was to lead the 1963 uprising, did not mince his words. "If it is wished that the Somali become sincere friends of the Abyssinians", he told the Emperor, "then the English subjects should be obliged to move out of the Ogaden immediately".[3]

The following year, the Italians invaded and occupied Ethiopia. In 1937, they concluded an agreement with the authorities of British Somaliland to allow the latter's subjects access to the Haud and the area between Jijiga and the Addis Ababa–Djibouti railway line which was called the Reserved Area. When the Italian East Africa Empire collapsed in 1941, and Ethiopia was liberated by British troops, the fate of the Ogaden hung in the balance. Claiming that it was essential for the prosecution of the war in the Far East, Britain retained control of the region, and forced an agreement to that effect on the reluctant Ethiopians; however, without prejudicing the latter's sovereign rights in the area. In fact, with Italy out of the Horn, Britain was planning to gather all the Somali under its wing. It seemed possible that Ethiopia could be compelled to trade the Ogaden for Eritrea, another former Italian colony and a far greater prize that was also under British rule at that time.

Throughout the 1940s, all the Somali lands, including the Northern Frontier District of Kenya, were under British rule. It was during this period that the Ishaq and other pastoralists from the British Protectorate in the north took advantage of colonial protection to make inroads into the Haud. They were driven by a worsening ecological situation, the first signs of which were noted at this time. The human population in the North was estimated at about 600,000, eighty-five percent of which were nomad pastoralists owning an estimated 1.2 million camels, 2.35 million sheep, 1.64 million goats and 223,000 cattle.[4] They were all served by deep wells. The area in the vicinity of the wells was denuded of vegetation and signs of degradation were evident elsewhere. One study found that the North had "a very high rate of overstocking",[5] and an official report in 1947 warned that "the soil and vegetation are on the brink of irreversible ruin" on account of congestion and overgrazing.[6] Congestion led to increased violence among the clans, and the Official Report for 1952–1953 cited a total of 80 deaths in disputes over grazing and watering rights.[7]

Under such ecological pressure, and with British forbearance, the nomads from the North moved en masse into the Haud and prolonged their stay there as long as possible. Fully half the population of the British colony with their animals came and stayed now up to nine months.[8] They nearly matched the human and animal population of the Ogaden itself. Lacking state protection, the latter were at a distinct disadvantage. Although Ethiopia recovered the Ogaden in 1948, Britain continued to hang on to the Haud and the Reserved Area, still hoping it might be possible to annex them.

A riot broke out in Jijiga when the flag of the Somali Youth League was lowered in July 1948, marking the restoration of Ethiopian rule. This was the first sign of militant Somali nationalism. The Somali Youth League was the first and foremost nationalist organisation, with an ethnic constituency that was predominantly Darod, the largest and most widespread Somali clan family. The Youth League was able to organise branches in all Somali regions, including the Ogaden, whose dominant clan belong to the Darod family. The Youth League was weakest in the British colony in the North. There, the Ishaq supported a rival nationalist organisation, the Somali National League, whose following was limited to this region. In this way, the Ishaq–Ogaden dispute was overlaid with a first layer of political antagonism that was not of their own making and was largely irrelevant to the material issue that was the core of their dispute. Both leagues had identical nationalist goals seeking independence and unity of all Somali territories. Significantly, the National League made no attempt to extend its activities across the border into the Ogaden, and there is no record of any protest by it when that region was returned to Ethiopia.

By contrast, there was a storm of protest in the North, orchestrated by the National League. when the Haud and the Reserved Area were finally handed back to Ethiopia in 1954. The British by now had given up hope of creating a pan-Somali Protectorate, partly because Italy had been granted a United Nations trusteeship over its former colony in the south in 1950, but mainly because the Somali nationalist organisations had expressed a clear preference for independence rather than any form of British tutelage. Both leagues organised protests against the return of the Haud and the Reserved Area to Ethiopia. The greatest exertion was made in the north where, inter alia, funds were collected to despatch two delegations of notables to carry the protest to the British government in London and the United Nations in New York.

To soften the blow, Britain secured an agreement with Ethiopia guaranteeing access to the Haud for its subjects for another fifteen years. Its implementation was to be supervised by liaison officers from the Protectorate, with the help of three hundred native police. Disputes between British subjects inside the Haud were to be adjudicated in the Protectorate, and various other provisions were included to safeguard the Ishaq and other northerners when they ventured into Ethiopian territory. The arrangement, in the words of the chief liaison officer, proved a 'fiasco'.[9] Understandably worried that it might serve as precedent for future Somali irredentist claims, the Ethiopians were concerned to limit the scope of the arrangement. Insisting that it applied only to nomads who stayed in the Haud for no more than six months per year, they refused to recognise as British protected subjects those who stayed longer or tried to cultivate grain as well. They also insisted on dealing with clans rather then individuals, and sought to persuade clan chiefs to opt for Ethiopian nationality, which was then held to apply to all the members of the clan. The Habr Awal, the largest Ishaq clan, some sections of which straddle the border, came under great Ethiopian pressure, and the tension was such that a journalist who visited the Haud at this time reported that "a state of war exists here".[10] Large-scale animal raiding contributed to the tension. The Ogaden now had the advantage, because the Ethiopian authorities saw to it that ani-

161

mals stolen from them were returned, but did not force the Ogaden to return animals they stole from the Ishaq. Finally, following a violent clash at Danod at the end of 1960, when the Ethiopians refused to allow herders from the British colony to draw water, Ethiopia abrogaged the agreement.

A new element was introduced into this volatile situation in the mid-1950s. This was the spectacular growth of the export market for animals in the oil rich Arab market across the narrow sea. Such exports had began in a modest way half a century earlier, when northern Somalia became the provider of meat for the British garrison at Aden. This was a minor item compared with the traditional export trade in hides, skins, ghee, gum arabic, myrrh and ivory. The expansion that occurs now is shown in Table no. 1.

The northern region provided around eighty-five percent of the animals exported. It is estimated that the Ogaden provided between twenty and forty percent of the animals exported through Berbera, the main northern port.

The pull of the market led to an appreciable increase in the size of the herds and flocks, especially sheep, the prime export item. To cope with the overload, cement water tanks called birket appeared in large numbers throughout Somalia. The idea, as well as the money to build them, came from Aden where many Somali find work. Equipped with petrol pumps, they rely mainly on stored rainwater, though sometimes they are filled with water brought by lorry. The first birket appeared in the Haud in 1956. By 1971, there were at least one thousand between Dibileh and Awareh. Nearly all were owned by Ogaden families. The Ishaq owned none and were obliged to buy water from the Ogaden in order to prolong their stay in the Haud. More of them began to stay all year clustered around the birkets, and the result of such congestion threatened to upset the delicate balance between land, water and animals that sustains traditional production. Congestion also led to a further increase in clashes among the clans frequenting the Haud. In December 1957, a dispute between Dolbahanda and Beidyhan left about one hundred and forty dead.

As noted above, the Ogaden were themselves involved in livestock export trade through northern Somalia. The attraction of that market, in contrast to Ethiopian interior, was a large difference in prices, with prices in Hargeisa being double those offered in Jijiga.[11] To reach its destination the Ogaden trade had to cross Ishaq territory, and the latter exacted heavy tolls to allow passage. In this way, the Ishaq compensated themselves for the price they paid to the Ogaden for water in the Haud.

The two former colonies merged in 1960 to form the independent Somali Republic. Three other Somali clans which inhabited territories–Ogaden, Djibouti, and the Northern Frontier District of Kenya remained outside the fold to become a source of perennial anguish and the apple of discord in the Horn. The merger of North and South has not produced a perfect union. With less than half the population of the South, and no sign of modern economic activity, northern Somalia became a junior partner in the state, a position it has found hard to accept. Resentment was initially manifested when the North rejected the proposed unitary constitution in the 1961 referendum. After the constitution was adopted with a large vote of support in the South, a group of junior officers in the North staged an abortive coup d'état. Since then the North has been a hotbed of political dissidence, spawning countless rebellions against the central government in Mogadisho. The Ishaq have been centrally involved in most of them.

By contrast, and despite the fact that they remained under Ethiopian rule, the Ogaden became closely linked with the ruling class in Somalia. The recovery of the Ogaden became a nationalist imperative binding on all Somali governments. It was put to a test in 1963, when a spontaneous uprising broke out in that region. The Somali army provided weapons, and the government in Mogadisho raised a diplomatic hue and cry on behalf of the rebels. Ethiopia's response was to attack Somali border posts, and the two countries drifted towards an undeclared war that lasted a few months. Somalia was soon forced to back down and cease its support for the rebels. When the uprising collapsed as a result, the Ogaden leadership and thousands of their followers sought refuge in the South, There they waited for another opportunity to challenge Ethiopia. Many joined the Somali army, where the Ogaden are well represented in the officer corps.

In 1969, a successful military coup d'état, led by General Mohammed Siad Barre, ended civilian rule in the Somali Republic. After an initial period of self-consolidation, the new regime resumed the quest for the recovery of lost Somali territories. The refugees from Ethiopian-held lands were regrouped in new organisations, some were sent to North Korea for guerrilla warfare training, and others were trained in Somali camps. As Ethiopia entered a period of political turmoil with the overthrow of Haile Selassie in September 1974, the temptation to wrest the Ogaden away from alien rule became irresistible. The organisation of Ogaden refugees was renamed the Western Somali Liberation Front, and its guerrilla units crossed into Ethiopia at the start of 1976, led by Ogaden officers of the Somali army.

The Western Somali Liberation Front recruited and armed Ogaden nomads, and commenced attacking Ethiopian outposts. Its ranks expanded quickly, and by the following spring they were besieging the region's administrative centres. The Ethiopian response was feeble. The army had become demoralised and disorganised by the widespread purges carried out by the embattled military junta in Addis Ababa. The

United States had suspended military assistance to Ethiopia, and the army was confronted with a serious shortage of ammunition and re- placement parts for its entirely American made weaponry. The Ethiopians had recently concluded an agreement with the Soviet Union that included military assistance. Until now Somalia's patron, the Soviet Union has now switched sides, with ominous consequences for Somalia. The latter's military rulers were confronted with a hard choice. Ethiopia's political turmoil and the poor performance of its army in the Ogaden presented a rare opportunity to reach for the pri- mary nationalist goal, i.e., the Ogaden. That opportunity had to be grasped immediately, before the Ethiopians had time to obtain and in- tegrate the promised Soviet weaponry.

Bowing to the imperative of Somali nationalism, the regime in Mo- gadisho committed its regular forces to an invasion of the Ogaden in the summer of 1977. They cleared the Ethiopian forces out of the region easily, and by early autumn they had laid siege to the town of Harar. The Ethiopian counter attack came in the early spring of 1978. Well supplied by the Russians and spearheaded by Cuban combat units, it quickly routed the Somali forces. In March, as the Ethiopian forces stood poising threat on the border, the Somali government sued for peace by declaring that its forces had withdrawn from the Ogaden. The Western Somali Liberation Front vowed to carry on the struggle, and it succeeded in maintaining scattered guerrilla units in the Ogaden, forc- ing the Cubans to mount a permanent guard.

These were hard years for the Ishaq. They were at odds with the military regime in Mogadisho from the outset. Ibrahim Egal, the Prime Minister removed by the coup d'état, was the first northerner to hold that post. The man who took over the reins of government, General Siad Barre, belongs to the Marehan clan, a branch of the Darod clan family, and his mother is an Ogaden. His response to the customary manifestations of political dissidence in the North was harsh, further alienating the people of that restless region. Moreover, the North was hit by drought in 1973–1975 and suffered heavy livestock losses. About one hundred thousand nomads had to be evacuated from the region, and many of them were resettled in other parts of the country.

The conflict in the Ogaden was another blow. The arming of the Ogaden nomads put the Ishaq of the Haud at great peril. They lost many lives and large number of animals. They blamed the Western Somali Liberation Front for siding with the Ogaden in herder disputes, and suspected that the Front encouraged its kinsmen in their efforts to evict the Ishaq from the Haud. On the national political scene, the Ogaden came to be regarded as one of the pillars of the military regime. Hostile clans, especially the Ishaq, derided the Western Somali Libera- tion Front as 'Siad's mercenaries'. Thus another layer of political antagonism was added to the old dispute between the Ishaq and the Ogaden, though this time it was not irrelevant to the material issue in the dispute.

Opposition to the military regime began to multiply and harden following their defeat in the Ogaden. Military morale was badly shaken and a series of mutinies and coups d'état were launched unsuccessfully in the years that followed. The survivors fled abroad, where they joined the crowd of political refugees produced by repression at home. In the late 1970s, two organisations were formed abroad to oppose the regime in Mogadisho. Typically, these had different regional and clan bases. The Somali National Movement represented mainly Ishaq dissidence in the North, and the Somali Salvation Democratic Front represented opposition in the South, especially among the Mijertein, a formerly prominent clan whose political fortunes suffered under military rule.

Ironically, both organisations solicited Ethiopian assistance, presenting the regime in Addis Ababa with an unhoped for opportunity to unravel the legend of Somali nationalism. Both groups set up bases near the border and armed by the Ethopians they began to raid Somali border posts and villages. The Somali National Movement has bases in the northern Ogaden, and its incursions are directed into northern Somalia. With local support there, its raids became bolder and went deeper into the North, and the insurgents were able to hold their own ground against the government forces. In the summer of 1988, the Somali National Movement forces were battling government troops for control of Hargeisa, the country's second largest town and capital of the North.

While battling the Somali regime, the rebels also waged another war against the Western Somali Liberation Front in the Ogaden, which they considered a tool of Siad Barre's regime. The Somali National Movement was in a position to intercept the Front's movements to and from Somalia, cutting off its supplies and communications. To the gratification of the Ethiopians, the Somali National Movement was able to do a much better job than they could in pursuing the scattered bands of nationalist guerrillas, and by the mid-1980s the Western Somali Liberation Front had no meaningful presence in the Ogaden.

While the dream of Somali nationalism was being shattered by a fractious ruling class, the pastoralists pursued their own feud single-mindedly. This time the Ishaq had the advantage, because the Somali National Movement recruited among them and put modern weapons in their hands. Not surprisingly, the weapons were turned against the Ogaden, who no longer had the protection of the Western Somali Liberation Front. Ogaden trade with Hargeisa and Berbera was disrupted also. Even travel there became hazardous, after a group of students returning to the Ogaden were killed by the Habr Yoonis section of the Ishaq in November 1982. The clan war that erupted then continues intermittently to this day. The Ogaden was devastated as a result. Squeezed between the Ethiopians on one side, and the Ishaq on the other, and menaced by recurrent drought, its people are drifting away from their homeland. Some are heading south to Somalia, others north to the settled areas of the Harar plateau. In either case, refugee camps are the final destination of the once proud and self-sufficient nomads.

*

The Ishaq–Ogaden conflict illustrates the persistence of age old material imperatives in the pastoralist world. This is not surprising, in view of the fact that the mode of production in this world has not changed, and the twin imperatives of extensive use of land and freedom of movement remain as valid as ever. Widespread participation in the market has affected pastoralist society in various ways, but it has not affected yet the traditional production technique; except in the case of the use of water tanks.

Defending these material imperatives has become increasingly difficult in the era of the modern state, and the pastoralists are often at odds with the states that claim them 'subjects'. In the Horn of Africa, the situation is further complicated by the intervention of various political movements fighting for sundry causes. The pastoralists are often entangled in such conflicts and appear to be fighting in tandem with other social groups for sentimental notions such as nationalism and religion and alien institutions like political parties and the state. Appearances are often deceiving. The commitment of pastoralists to such causes and the movements that represent them is normally conditioned by expediency. This might be as simple as an opportunity to obtain modern arms for the defence of their herds and territory, or a more complex yet easily understandable impulse to weaken the all-encompassing authority of the modern state. Pastoralists in the Horn have proved enthusiastic recruits for movements whose goal is to break up the present state structures in the region.[12]

NOTES

1. Cited in Said Samatar, 1982: *Oral Poetry and Somali Nationalism*. Cambridge University Press, p. 19.
2. Lewis. I.M. 1969: *Peoples of the Horn of Africa*. London, International African Institute, p. 107.
3. Lessona, A. 1939: *Verso l'Impero*. Firenze, p. 100.
4. Silberman, L. 1959: "Somali Nomads" in *International Sociological Science Journal*, 11,4 p. 571.
5. Gililand, H.B. 1952: "The Vegetation of.Eastern British Somaliland". *Journal of Ecology*, 40,2, p. 94.
6. Cited in Geshekter, C.L. 1983: "Anti-Colonialism and Class Formation: The Horn of Africa". Paper presented to the Second International Congress of Somali Studies, Hamburg, p. 23.
7. Colonial Report, *British Somaliland Protectorate 1952, 1953*, p. 3.
8. Drysdale, J. 1964: *The Somali Dispute*. London, Pall Mall Press, p. 79
9. Drysdale, 1964: chapter 7.
10. *The Times*. London. 27 October 1956.
11. Swift, J. 1979: "The Development of Livestock Trading in a Nomad pastoralist Society". In Equipe Ecologie et Anthropologie des Societés pastorales, *Pastoral Production and Society*. Cambridge University Press, p. 452.
12. Cossins, N. 1971: "Pastoralism Under Pressure: A Study of the Somali Clans of the Jijiga area of Ethiopia". Meat and Livestock Board, Addis Ababa, p. 90.
13. For a discussion of this phenomenon see Markakis, J. 1987: *National and Class Conflict in the Horn of Africa*. Cambridge University Press.

TABLES AND MAP

Table I. *Somalia: Export of livestock*

	1950	1959	1963	1972
Sheep & Goats (in thousands)	121	455	829	1,636
Camels (head)	174	3,613	15,302	21,954
Cattle (in thousands)	2.7	14.2	40	81.3

Source: Swift J, "The Development of Livestock Trading in a Nomad Pastoralist Society", in Equipe Ecologie et Anthropologie des Societés Pastorales, *Pastoral Production and Society* (Cambridge University Press, 1979) p. 452.

Map I. *The Ogaden*

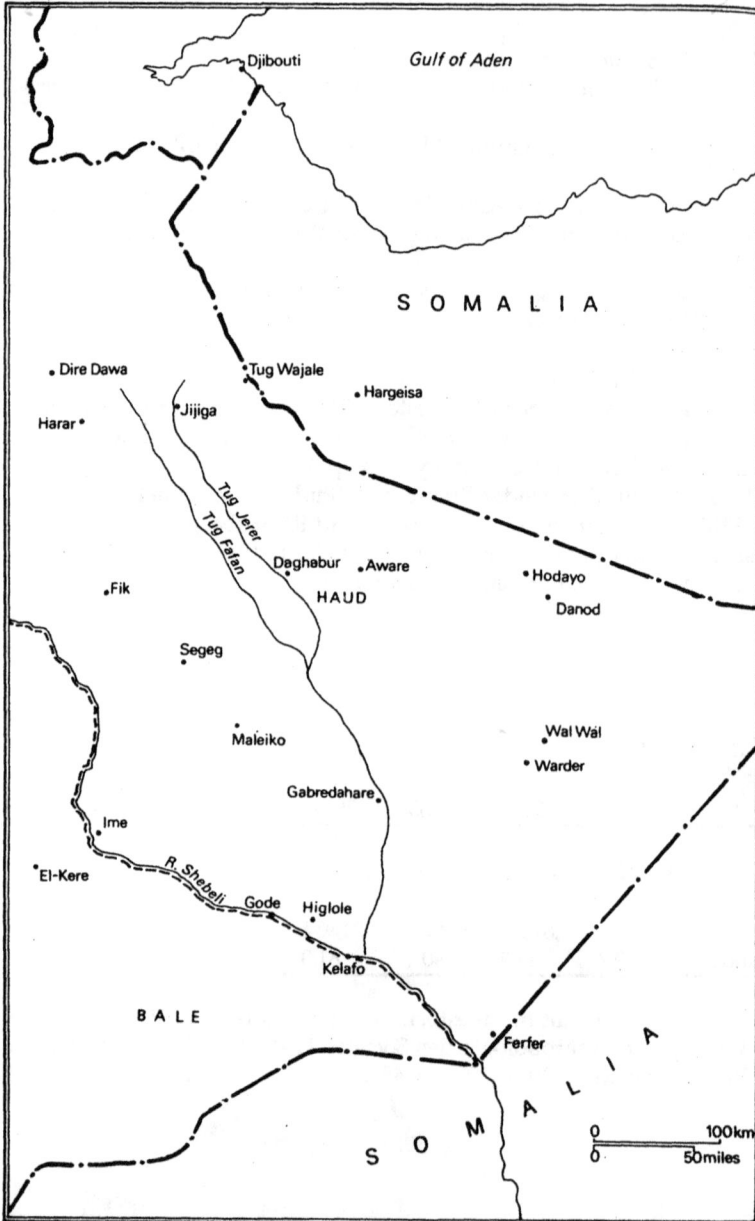

Source: Markakis, J., *National and Class Conflict in the Horn of Africa*, 1987.

Desertification, Refugees and Regional Conflict in West Africa

Okwudiba Nnoli

DESERTIFICATION IN THE WEST AFRICAN REGION

Desertification has become a major problem of the African States. The drought that plagued the African continent between 1972 and 1987 has caused the soil to dry up in many countries, the water table to be lowered, the vegetation to vanish and life in general to become extremely hazardous. For example, Africa has lost an average of 36,000 square kilometres of forest to the desert every year. In 1980 alone, 200,000 square kilometres of arable land were lost.[1]

Consequently, the continent needs $80 billion to fight desertification in the next 20 years. In other words, it requires $4 billion every year for the next 20 years for this purpose.[2]

The situation is particularly serious in the Sahelian zone of Africa in which many West African countries are located. Notable among these countries and areas are Niger, Mali, Burkina Faso, Mauritania, Northern Nigeria, Northern Ghana, Senegal, Mauritania, and parts of Gambia. Of the six hardest hit African nations, Chad, Ethiopia, Mali Mauritania, Niger and Sudan, three are West African and one, Chad, straddles West and Central Africa. The Sahel countries have experienced continued rainfall shortages over the fast fifteen years. In 1983 and 1984 they recorded their lowest total in a century. The resultant drought has accelerated the deterioration of the land leading to desertification.[3]

In fact, the Sahara desert is slowly but steadily advancing on these countries. For example, between 1972 and 1985 there was no substantial rain in Mauritania. The water table dropped making it more costly to drill wells. Whatever natural vegetation that remained was eaten by hungry cattle. Deprived of their stabilizing vegetation sand dunes have continued to shift, and the desert to advance by some twenty kilometres every year.[4] Sandstorms occur almost every other day affecting Africans as far south as the Senegal river and sometimes even Dakar. In Nouakchott, Mauritania's capital, sand storms rage so fiercely that at midday residents must use headlights to see the road.

As the desert advances, river beds dry up making irrigation impossible. This development together with the lowering of the water table and the continuing drought renders agricultural production untenable. Drought, by thus immediately affecting agricultural production, sharpens the competition between export crops and food crops for land, family labour inputs and other available resources. Prolonged drought undermines the adaptive and adjustive mechanisms of the people and the ecology, leading to famine.

Since nearly 80% of the African population live in the rural areas and make their living by farming, famine creates havoc for the vast majority of the population and drastic consequences for the rest of the society. United Nations agencies estimate that in 1985 famine in Africa placed between 30 and 35 million people in 20 countries seriously at risk. Almost 10 million people had to abandon their homes and lands in search of food and water.[5] Hundreds of thousands of famine-stricken people are departing the countryside; they face a bleak future in shanty towns of already crowded cities, and an uncertain future in neighbouring countries.

In Mali for example, a total of 1.2 million people were affected by drought, famine and related diseases in 1985. Losses of animals and crops drove 200,000 people toward major population centres, especially those of Timbouctou, Gao and Mopti along the Niger River. Not only were the people plagued by drought, poor health left them vulnerable to cholera and other diseases. Between May and August 1985 cholera alone struck 1,610 people and claimed 286 lives.[6] Formerly, about 80% of Mauritania's 1.7 million people were nomadic cattle herders. Lack of water and grazing land is depleting their herds and driving most of them into shanty towns. This population shift to towns creates social and economic problems. Traditional homelands are abandoned, and large groups of people dependent on external aid crowd into relief centres, creating an environmental nightmare.

However, this famine-generated migration is not confined to the towns in the drought-stricken country. Drought victims also migrate to rural areas within and across the territory; from urban to urban to urban areas, and from urban to rural areas both internally and internationally. It is different from the normal migration, which also follows these patterns, because families and entire communities are forced to migrate in search of food. Although these migrations begin as temporary measures they may become permanent, depending on the recovery of agriculture in the post-famine years. Furthermore, such migration is often accompanied by expanding hunger and increased infant mortality, deterioration of the socio-economic infrastructure at the end point of migration, shortages of consumer goods and high rates of inflation at the destination, and increased corruption and internal social conflict both within the migrant community and with the host community.

Thus drought-related migration is potentially a source of conflict both within and across nations. The logic is simple. It relates to the link between peace and development. It is well known that unless there is peace, development is not possible. What is not as well known is that unless there is development, peace is not possible and if it exists will not last. Through development individuals and groups are able to achieve security by rising above the worst forms of human degradation and suffering such as are imposed by hunger disease, illiteracy and sheer drudgery. Such conditions provide a fertile ground for the eruption of conflict. The degradation of refugee life caused by desertification provides such a condition. Often the victims of drought and famine face the threat of mass starvation, and the reality of mass starvation. Men, women and children have craggy skins, thin limbs and protuberous abdomens. They are victims of hunger and preventable diseases that derive from malnutrition. Inside the dwelling place they are herded together under conditions which defy the rules of hygiene. Psychologically, they must suffer the humiliation of dependence for their very existence on relief materials provided by humanitarian organisations. This is a situation of structural violence which in one form or another, sooner or later explodes into open conflict.

FAMINE-REFUGEES AND THE FLUID BOUNDARIES OF WEST AFRICA

As a result of desertification in the West African region, famine-refugees have spilled across various state boundaries, creating assorted problems not only for their host countries but also in relations between their countries of origin and the host nations. In the Ivory Coast such victims from the Sahel region to the north of the country have found refuge. Peoples from Mali, Burkina Faso and Niger migrate there to take advantage of better opportunities for life than in their own drought-stricken countries. Nigeria not only has to battle with desertification and its victims in its northern states of Sokoto, Katsina, Kano and Borno, it has also to contend with similar victims from Niger, Chad and Mali. Similarly, Ghana must grapple with the effects of desertification in northern Ghana as well as the victims of drought from Burkina Faso, Mali and Niger. Senegal must cope with such victims from its own territory as well as from Mauritania to the north.

Part of the inter-state conflict that has arisen in West Africa over famine-refugees is related to the largely indeterminate nature of inter-state boundaries in the region. State boundaries established by the erstwhile colonial powers were drawn "across well-established lines of communication including, in every case, a dormant or active sense of community based traditions concerning common ancestry, usually very strong kinship ties, shared socio-political institutions and economic resources, common customs and practices, and sometimes acceptance of a common political control."[7] In some cases the boundaries have separated communities of worshippers from age-old sacred groves and shrines. In some other instances even the water resources of a pastoral and nomadic people were located in one state while the pastures were in another.[8]

The Gourma are divided into parts located in Burkina Faso, Togo and Benin. The Wolof and Sere live across the Senegal and Gambia boundaries; so do the Soninke and Tukulor of Mauritania and Senegal, the Hausa-Fulani of Niger and Nigeria; and the Akan of Ghana, Ivory Coast and Burkina Faso. Given the social fluidity of the boundaries many government authorities within states do not know precisely where their state boundaries are located. In any case, partitioned Africans have tended to ignore the boundaries as dividing lines in their normal activities and to carry on social relations across them more or less as in the days before the partition.

Many of the victims of desertification move across the borders and organise a new life for themselves among their kith and kin on the other side whenever the situation permits. A more serious problem for the refugees arises when their kith and kin are also victims of the drought. Historically, this famine-related migration of whole communities generated wars. For example, before 1880 the Senegambian border region was characterized by a high incidence of wars of politico-economic origin. These wars were based on competition for scarce resources such as arable or pastoral land, as well as for the control of flourishing trading centres and trade routes. In the battles of survival much enmity was generated among the ethnic groups of the region.[9]

Today such migrations complicate the problems of inter-state border control and management. The most serious problems concern an increase in crime and smuggling. Whether real or imagined the states involved tend to perceive an increase in crime and smuggling as attendant on such migrations. An attempt to control these anti-social activities many times leads to conflict among the states. The different administrative, monetary, and economic systems across the borders encourage these activities. The latter, therefore, infringe on the sovereignty of the states. The ethnography of the West African inter-state boundaries poses a permanent challenge to the role of the modern African state institutions charged with the task of screening the movement of persons and materials into the areas of jurisdiction of the relevant states.

Smuggling of export crops is the most sensitive of the problems that are generated by uncontrolled migration such as famine-related migration. But it should be seen as part of a wider network of activities featuring not only illegal trade in export goods but all forms of unauthorized movement of persons across state boundaries including even criminals who seek sanctuary from laws on the other side of the boundary. Such smuggling is rife in the Nigeria–Niger border region where Hausa traders are constantly in search of higher prices for their groundnut and other crops, Similarly, the Wolof and Sere smuggle groundnut across the Senegal–Gambia border. These illegal activities go on under normal circumstances. but they are intensified by the uncontrolled nature of famine-related migration. In addition such migrations make the control of these activities difficult. They disrupt the law-enforcement mechanism and generate bad blood between the neighbouring states because of the threat they pose to the various countries' competitive price advantage. This threat is compounded by drought-related West African groundnut shortages that have kept the world market prices for the crop high. These high prices plus the renewed strength of the CFA franc relative to the Naira in border currency markets, enabled Niger to meet and even beat the producer price increases in Nigeria.

INTER-STATE RESPONSE TO FAMINE-REFUGEES

Socially, desertification and the consequent migration of whole communities have created a number of problems. One of these is the social nuisance caused by beggars in countries or regions of a country where begging culture does not exist. As a result of the severe deprivation caused by famine, many famine-refugees rely on begging to survive, even when they have been accommodated by their kith and kin across the border. Often they migrate further south where begging may be regarded as a taboo. Here the begging migrants can only constitute a social nuisance. In certain parts of Southern Nigeria tough government actions were taken to control the situation. This created difficulties between Nigeria on the one hand and Niger and Chad from where the migrants originated on the other.

A much more serious problem arises from competition for the very scarce social and economic resources of the host country between the migrants and the host population. This is exacerbated by the fluidity of state boundaries which enables famine refugees to pass as the host population rather than concentrate in refugee relief centres. Under normal circumstances the socio-economic condition of the African country is very difficult. In the West African region the per capita income is approximately the same as the per capita national debt. The rate of inflation is up to 25% annually. It has a very high infant mortality of about 146 per thousand, and a low life expectancy of 47 years. In the urban areas there is one doctor per 700 of population, and in the rural areas one per 26 thousand. Less than 50% of the eligible children attend primary school and less than 30% attend secondary school. Worse still, whereas the average food growth rate of Africa was 2.5% between 1960 and 1979 it had fallen to 1.7% during the period 1970–1980. If the rate of growth of population is taken into account this rate of growth of food production was 0.7% between 1960 and 1970 and 0.9% between 1970 and 1982.[10]

Therefore, the social and economic infrastructure of the region is very fragile indeed. The influx of thousands of migrants largely in an uncontrolled manner stretches it beyond the limits. In fact, without famine relief assistance from outside Africa the situation would have been catastrophic. One of the ways in which host countries have responded to the situation is to treat the migrants who are not formal refugees as aliens who are required to register their presence with the law enforcement agencies under laws and conditions concerning alien residents. These migrants are subsequently expelled back to their country of origin when the legal term for their stay expires. This mass return of the migrants together with the suddeness and harsh conditions of their expulsion generates conflict between their home countries and their host country.

The mass expulsions of aliens from Nigeria in 1983 and 1985 is illustrative. Although all the aliens expelled were not drought-related refugees a good number of them were. In February 1983, the Nigerian government gave the largely non-Nigerian West African population who were residing in the country without valid formal immigration papers two weeks to leave the country. The time limit was later extended to one month for the benefit of the skilled workers. The resultant chaos and suffering by these expellees, which was compounded by the closure of the Ghana–Togo border at the time, generated bad blood between Nigeria and many West African states, and attracted criticism from within Nigeria itself. Those affected included famine refugees from Niger, Chad and Cameroon. However, the largest number came from Ghana and most of them were not drought victims. The exercise was repeated in 1985 but on a reduced scale. Nevertheless, the interstate political repercussion were the same. Associated with this expulsion is the complete sealing off of the Nigerian borders between September 1982 and March 1983.

This Nigerian action is not an isolated one in the West African region. As far back as December 1969, Ghana under Busia expelled over 1000 Nigerians as illegal aliens; in 1982 Sierra Leone expelled Guineans. Also, between September 1982 and March 1983 Ghana closed her borders. However, these non-Nigerian examples did not involve famine-refugees. Only the Nigerian expulsion involved them. In addition, the numbers expelled were staggering.

The official Nigerian reason for the 1983 expulsion included the accusation that the aliens were responsible for the high rate of crime in the country, and the high rate of unemployment in the nation because they displaced Nigerians from their jobs by their acceptance of lower wages than their Nigerian counterparts. The Nigerian Central Bank report for 1982/83 showed unemployment as 59.4 % of the relevant population. The aliens, particularly the famine-refugees from Niger, Chad and Cameroon, were also officially accused of involvement in the Maitatsine religious disturbances in Kano, Kaduna, and Maiduguri in Northern Nigeria. However, the real cause for the expulsion is the deepening economic crisis of the country which had been intensifying since early 1982.[11]

The rapid Nigerian economic growth of the 1970s began in 1979 to be replaced by stagnation and recession. By 1985 the GDP had fallen to the 1975 level when it should have risen by at least 7.5% per annum between 1980 and 1984 as envisaged by official planning instruments. Foreign exchange from the country's major export product, petroleum, fell dramatically to NGN 7.5 billion in 1983 from the 1980 figure of 13.5 billion. The external debt rose from NGN 30 million in 1973 to a staggering 20 billion in 1985. Inflation was running at 35% and the rising unemployment of school leavers and university graduates had raised the unemployment rate to 5% from the practically full employment of the early 1970s. This dire economic condition gave rise to the harsh Economic Stabilization Act of 1982 which imposed severe austerity measures on the population. The expulsion of aliens was undertaken in part to win back the popular support lost by the ruling party following these austerity measures as the 1983 elections drew near.[12]

These expulsions are particularly illuminating because they contradict the spirit of the Treaty establishing the Economic Community of West African States (ECOWAS). The treaty provides for the Free Movement of Persons, Residence and Establishment in Articles 2 (2d) and 27.[13] Article 2 (2d) states that "The community shall by stages ensure the abolition as between the Member States of the obstacles to the free movement of persons services and capital. Article 27(1) confers the status of community citizenship to citizens of member states and enjoins the latter to abolish all obstacles to their free movement, residence and establishment within the community. However, the treaty did not spell out in detail the stages to be followed and the time-table for the necessary mutual agreement on implementation. These were filled in by the Protocol later annexed to the treaty.[14] Although from the point of view of the Protocol relating to the Free Movement of Persons, Residence and Establishment, the Nigerian expulsions of 1983 and 1985 did not violate the letter of the Treaty they nevertheless violated the spirit of it.

RESTRAINTS ON INTER-STATE CONFLICT OVER FAMINE-REFUGEES

Fortunately, however, the potential for inter-state conflict inherent in cross-border migration of famine-refugees arising from desertification in West Africa has not been realized. This is explained in part by the comparatively high level of humaneness and compassion with which refugees are received in Africa compared to the other parts of the world. The refugee literature explains it as a demonstration of Africa's traditional hospitality. Also, the fact that the people who flee across state borders are welcomed on the other side by their kith and kin who may have an influence on national refugee policy is a factor.[15] In addition, sentiments of Pan African unity and solidarity bequeathed by Kwame Nkrumah to his friends and foes alike, as well as to posterity, in Africa have endured beyond the imaginations of many. This is why expulsions of fellow Africans for non-criminal reasons are often decried by most Africans, including citizens of the subject country. In the particular case of West Africa, the spirit of ECOWAS is fast pervading the inter-state political atmosphere. It is reflected in the Treaty relating to the Free Movement of Persons, Residence and Establishments, the abolition of visa requirements for West Africans travelling in the Region, and in the increasing political will for inter-state coordination and cooperation.

In another respect, desertification creates conflicts among African states that are not related to famine-refugees. Such conflicts are more directly related to desertification itself than through refugees. They concern the control of rivers which flow through the various countries. As drought intensifies there is a great temptation on the part of the drought-stricken country to dam the rivers in order to generate resources for irrigation purposes. Such dams disturb the normal flow of water to countries below the dam thereby adversely affecting their own ecology and their ability to continue to exact a living by their age-old reliance on the water of the affected rivers. The result is conflict between the country that dams the river and its neighbours downstream.

In 1977 for example, Niger constructed two dams on Rivers Lamido and Maggiya. These constructions disrupted irrigation projects along the River Kamano in Sokoto State of Northern Nigeria into which the rivers flow. This dam construction generated conflict between Niger and Nigeria.[16] A similar conflict has been generated between Nigeria and Cameroon. The latter has constructed the Laddo dam on the upstream of River Benue which has caused serious havoc to fishing, livestock and agricultural activities in the Gongola State of Nigeria.[17] It is in order to avoid such a conflict that Nigeria agreed to supply electricity to Niger from the former's Kainji dam power station. This dam was constructed on River Niger well below the Niger–Nigeria border. However, the supply of electricity to Niger removes the need for the latter to provide itself with electricity by building a dam on the same river that will disrupt economic activities in those areas of Nigeria close to the river. Also, a dispute is raging at the moment between Nigeria, Chad and Cameroon over the flow of rivers from their respective territories into Lake Chad, The drought in Nigeria and Cameroon has led to their daming of some rivers that traditionally flow into the lake. Consequently, the lake has been receding and its ecology is being disrupted. This has caused economic dislocation to large populations especially on the Nigerian side of the lake. Discussions are still going on to resolve the conflict.

In order to anticipate and diffuse conflicts likely to emanate from this control of inter-state rivers the West African states have created various river and lake commissions. Among these are the River Niger Commission and the Lake Chad Commission. The task of these commissions is to regulate and coordinate the activities of the member states regarding the use of the rivers which traverse their various territories or lakes whose waters extend across their respective borders. The objective is to ensure that no state suffers unduly as a result of the activities of one or the other states located near the water resource. Without the activities of the commissions the incidence of inter-state conflicts over the use of these resources would have been much greater in number than what it is now.

THE NEED FOR LEGAL PROTECTION OF FAMINE-REFUGEES

In conclusion, it is clear that an explosive situation exists in West Africa arising from the cross border migration of refugees fleeing from desertification, and from the conflicting interests of states over inter-state rivers and lakes as they battle against the adverse economic effects of desertification. At the moment this danger has been averted because of the persistence of African solidarity, hospitality and effective coordination of inter-state policies on the use of inter-state rivers and lakes. However, the growing impoverishment and marginalisation of communities and regions that host the famine-refugees, and the drought-stricken countries themselves, severely undermine the capacity of the West African states to continue in the path of peaceful resolution of problems generated by desertification.

Ultimately the question of building the capacity of the African states is one of confronting imperialism, both its internal and external aspects. Internally a self-serving neo-colonial ruling circle focuses essentially on what benefits it derives from the neo-colonial scheme of things. Increasingly divorced from its people it cannot articulate their interests or pursue them without wavering. In the face of the declining economic conditions of the society the leader finds himself helpless. He is hostage to the accumulated national debt, the whims and caprices of foreign capitalist creditors, the IMF and the World Bank, and the economic stranglehold of the advanced capitalist societies generally. His high-handed, administrative, inefficient and corrupt approach to governance clearly contradicts his people's material and democratic aspirations.

Externally, imperialism is increasingly imposing the burden for progress out of the world-wide crisis of capitalism on the masses in African-type societies. Subsidies are removed, currencies are devalued and employment is severely restricted. The people get poorer and poorer. Even the so-called middle class is pauperized as the society is polarized into the few rich and many poor. In West Africa, the 1980s have been characterized by higher prices of exports. The result is a balance of payments crisis. And without foreign exchange reserves, suppliers' credits dry up. At the same time the North–South Dialogue is discontinued, a New International Economic Order is rejected, the IMF and World Bank remain uncompromising in their policies, UNCTAD conferences cannot be productive, and the inexorability of the debt trap persists. Under these conditions the people become more hungry and miserable, they cannot resist desertification or cope with famine. Their leaders, in their helplessness and mismanagement, will not be able to contain the consequent conflicts among them either internally or in the West African inter-state system.

Thus in the absence of any immediate prospects for ameliorating the consequences of desertification it is necessary at least to ensure the protection of famine-related refugees. In this regard, despite the great strides made in refugee law in Africa by the 1969 OAU Convention Governing the Specific Aspects of Refugees Problems in Africa, and the 1981 African Charter on Human and Peoples' Rights, African refugee law does.not address the situation of persons displaced as a result of drought and famine especially when they do not formally assume the status of refugee. Such protection should include the rights of individuals, groups and communities stricken by famine to cross international boundaries in search of succour, and to be treated with humanitarian considerations until the cause of their migration has been removed and the conditions are favourable for their return to their home country. Meanwhile the struggle against desertification must continue.

NOTES

1. Cf. A report of the commission set up by the European Community on the problem of desertification as reported by *The Republic,,* Monday, August 22, 1988, p. 12.
2. *Ibid.*
3. Michael J Schultheis, "A Continent in Crisis: Migrants and Refugees in Africa", paper presented at the conference on The African Context of Human Rights, University of Port Harcourt, Nigeria, June 9–11, 1987, p. 3.
4. "Update on UNDRO'S Activities in Africa" *UNDRO NEWS.* July/August 1985, p. 8.
5. UN Office for emergency Operations in Africa, Special Report on the Emergency Situation in Africa: *Review of 1985 and 1986 Emergency Needs* (New York, UN January 30th 1986).
6. *UNDRO NEWS, op. cit.,* p. 7.
7. A.I. Asiwaju, ed., *Partitioned Africans* (London: Hurst, 1984) p. 2.
8. *Ibid.,* p. 3.
9. *Ibid.,* p. 74.
10. These figures have been taken from the following sources: R.L. Siverd, *World Military and Social Expenditures 1987–88* (Washington D.C: World Priorities, 1987); Colin Legum et al, *Africa in the 1980s: A Continent in Crisis* (NY: McGraw-Hill, 1979); Thea Buttner and Hans-Ulrich Walter, *Colonialism, Neo-colonialism and the Anti-Imperialist Struggle in Africa* (Berlin: Akademie-Verlag, 1984); Lothar Rothman, "Colonialism, Neo-colonialism and Africa's Path to a Peaceful Future. Historical, Political and Economic Aspects".Paper presented at the centenary conference of the Berlin Congress of 1884–85 held in Berlin, GDR, February 6–8, 1985; "Peace Research and disarmament in Africa" A Gyor Conference Document in *International Peace Research Newsletter,* Vol. XXII, No.4, 1984, p. 3; The World Bank, *Toward Sustained Development in Sub-Saharan Africa* Washington D.C.: World Bank, 1984.
11. On the 1983 and 1985 expulsion of aliens refer to the following: Isaac Aluko-Olokan, "The Nigerian Economy in Crisis", *Nigerian Forum,* Vol. 2, No. 6, June 1982, pp. 584–595; Femi Aribisala, "ECOWAS, Past, Present and Future", *Nigerian Forum,* Vol. 5, Nos 5 and 6, May/June 1985; S.O. Ojiako, BA research project Department of Political Science, University of Nigeria, Nsukka.
12. R. Omotayo Olaniyan, "Nigeria and the Economic Community of West African States: A Role and Problem Analysis" in G.O. Olusanya and R.A. Akindele, eds., *Nigeria External Relations: The First Twenty-Five Years* (Ibadan: University Press, 1986) p. 132.
13. Treaty of the Economic Community of West African States (ECOWAS) (Lagos: ECOWAS Secretariat, 1975).
14. In Nigeria, for example, it is believed that some high ranking members of the Armed Forces are from Niger and Chad.
15. Oscar Ede, "Nigeria and Francophone Africa" in /G.O. Clusanya and R.A. Akindele, eds., *op. cit.* p. 185.
16. *Ibid.,* p. 186.

Environmental Degradation and Political Constraints in Ethiopia

Michael Ståhl

INTRODUCTION

Environmental degradation in Ethiopia is well documented. It is estimated that 50% of the highlands are significantly eroded, while 25% are seriously eroded and 4% are beyond recovery.[1] If present trends in grazing, cultivation and removal of vegetation continue, the physical resource base in the northern highlands will be overtaken by irreversible degradation at the turn of the century.

Ethiopia's political situation shows disparate tendencies. In the central regions, which also are the surplus producing agricultural areas, the political situation is stable. Government is in firm control and large-scale socialist transformation programmes are implemented. By contrast, the northern highlands where environmental degradation is worst, are torn by civil war.

In this paper it is argued that the soil conservation and related environmental rehabilitation programmes presently implemented by the government with donor assistance, will not be able to reverse the trend to environmental degradation. Their scope and funding is too limited. At the same time, the incentive structure for peasants participating in conservation activities is insufficient. It is argued that the low priority given to environmental rehabilitation by the government, peasants and donors are ultimately related to the unresolved political problems in norhtern Ethiopia, which consume the bulk of available resources.

The analysis refers primarily to the mid-1980s and examples are taken from Wollo region.

ENVIRONMENTAL DEGRADATION

The highlands of northern Ethiopia were once covered by large forests. Their volcanic soils were fertile. The highlands were a centre of plant genetic diversity. The cool climate attracted early human settlement. Mixed agriculture and stock keeping was established several thousand years ago.[2] Plough agriculture and irrigation emerged and provided,

181

together with long distance trade, the economic foundations for political centralization.[3]

The great majority of the rural people live in the highlands (above 1,500 m) where the climate is sub-tropical, rainfall is adequate in "normal" years and soils are fertile. These highlands are the traditional centres of population and livestock concentration and have been exposed to cultivation and grazing for centuries, not to say millenia. Historically, the political centre has moved from the north southwards. The ancient Axumite kingdom, which flourished in the first centuries A.D., was located in the highlands of what presently are the regions of Eritrea and Tigray. Its successor, the Zagwe kingdom (1000–1300 A.D.) was located in northern Wollo while the Abyssinian kingdom, which emerged in the 14th century, had its centres in the present day regions of Gondar and Shoa, slowly penetrating to the south. Addis Ababa was founded in 1886, some 800 km south of Axum.[4]

The drift to the south has many explanations. One is ecological. The highland soils are erodible and degradation sets in when pressure of land use exceeds a certain level, i.e. when trees have been felled, vegetation intensively grazed and the soil ploughed and left bare during the heavy rains. The hilly topography and the high intensity of rainfall increases the rate of erosion. Traditionally the peasants cleared virgin land once yields started to diminish due to degradation.

The expansion southwards was usually organized as military campaigns by the Abyssinian king against Muslim and pagan communities. With the influx of Christian settlers in the new territories, a farming system based on cereal crops, draught oxen and ploughs was introduced.[5]

Population growth in the 20th century, enhanced by partial control of epidemics and by the relatively peaceful period of Haile Selassie's rule after the second world war, has given land pressure a new dimension. Population growth is now estimated at 2.9% per annum.[6] In the northern parts of the highlands, the scope for further areal expansion of cropland is negligible. There is no virgin land any longer. Hence cultivation becomes more intensive. Long fallow periods cannot be upheld. The soil is cultivated every year. Soil fertilization through organic manuring has become less frequent because the scarcity of firewood forces people to use cow-dung as household fuel. A survey of an area above 2,500 m in Wollo carried out in 1987 found that cow dung constituted 80% of household fuel. In the same area, local grazing was sufficient for less than 50% of the required feed intake for efficient draught and milk production.[7]

Today one can divide the highlands of Ethiopia into two main parts. In the "south" (south of Addis Ababa) there is still woody vegetation left. Soils are relatively deep and fertile and rainfall is usually adequate for arable farming. Given appropriate agricultural policies, surpluses from crop and animal husbandry could be produced. In the "north" the landscape is generally barren. After centuries of exploitation the forests have been cleared. Shallow soils cannot retain rainwater, much of which disappears as runoff. Both sheet wash and gullies threaten farmland.

This differentiation of "north" and "south" is not entirely valid. There are fertile and densely vegetated areas north of Addis Ababa while some southern areas display severe land degradation.

A detailed study of the land degradation process in Ethiopia was carried out in the mid 1980s.[8] The reports show that erosion is heaviest on agricultural land which lies bare at the onset of the summer rains. If erosion proceeds unabatedly, one fourth of the present arable land in the highlands may be unable to sustain crop production by 2010. Parts of Eritrea and Tigray have already been so badly eroded that they are barren. Erosion has decreased in such areas because most soil is already gone. The regions of Gondar, Wollo and northern Shoa are now in the frontline of environmental degradation. Deep soils and vegetation are still left in many places but the intensive cultivation and grazing has reduced the soil waterholding capacity and makes them prone to drought.

As a consequence of environmental degradation, land has progressively shifted to less productive uses—crop yields decrease, the nutritional composition of grasses has deteriorated, providing poor grazing for livestock and, eventually, the land produces neither crops nor feed for cattle.

At the same time, the continuing population increase linked to diminishing productive resources compounds the imbalance between actual agricultural production and consumption needs. In parts of Wollo per capita consumption is now down to 60–70% of WHO recommended calory intake.[9] Seasonal hunger is chronic. Case studies from Wollo show that food availability is lowest during June to September. This is the rainy season when hygienic conditions around homesteads get increasingly polluted and water-related diseases are at a peak. Diarrhoeal diseases have deleterious effects on both children and adults and many working days are lost. This is also the time when heavy labour is required in seedbed preparation, planting and weeding.[10] A sick and exhausted peasantry faces the task of obtaining a livelihood from the dwindling physical resource base.

In addition to feeding themselves, the peasants must feed the public sector. The state procurement agency, the Agricultural Marketing Corporation—AMC, is empowered to procure a quota of grain from every peasant household at a fixed price.[11] The quota system is firmly established through the peasant associations and supported by the local administration. Quotas vary from year to year and are set differently according to the estimated production capacity of the local community. Although the quota for Wollo was lowered during the drought 1984, it was not abolished.[12]

When drought comes, peasants' reserves are rapidly exhausted and the stage is set for famine. The northern highlands have experienced six serious famines during this century. The devastating effects of the 1984/85 drought and famine have been commendably analyzed in a recent monograph on Wollo.[13]

Although acute disasters are related to drought, the decreasing ability of peasant communities to feed themselves is related to population growth and to the lack of technology improvement. The highlanders use oxen-drawn ploughs for soil preparation. Most other operations are done by hand using hoes and sickles. This technology was well adapted to the agricultural circumstances when long fallow systems could be upheld. However, peasants are now obliged to keep the fields in continuous cultivation or grazing. Under intensive land use the traditional agricultural technology encourages erosion. The sloping agricultural fields lie unprotected when the early rainstorms hit the highlands.

Nobody can pinpoint when population growth in northern Ethiopia "took off". The lack of vital statistics as well as general demographic and socio-economic information from the area makes this task difficult. Sometime during this century population and livestock growth appear to have reached a threshold where the local ecology could no longer cope. This happened less than a generation ago. It is instructive to drive through the northern highlands with elderly townspeople originating in the area. They remember a "jungle", "thick forest", "impenetrable bush" in places where today most vegetation is gone.

ENVIRONMENTAL REHABILITATION PROGRAMMES

Activities and organization

The Ministry of Agriculture (MOA) classifies the rural areas of Ethiopia into high potential surplus-producing zones and low potential subsistence zones. Intensive development programmes (seed-fertilizer packages) are emphasized in the former zones which include the well watered highland plateaux of Shoa, Arsi, Bale and Gojam.*

However, only one-third of the administrative districts are classified as having a high agricultural potential. The subsistence areas include the eroded northern highlands. They have low priority in terms of government funding for agriculture and staff deployment. In Wollo, only two of the twelve administrative districts are classified as food surplus areas.[14]

Nevertheless, programmes have been designed by MOA for rehabilitation of the eroded areas. Activities are concentrated to Wollo, Gondar and northern Shoa as well as to Hararge in the east. In Eritrea and Tigray, rehabilitation programmes are implemented at a reduced level or not at all due to the unstable political and military situation. The activities will be briefly presented below.

The activities promoted by MOA concentrate on terracing, closing of hillsides and tree-planting. The following brief survey relates to Wollo.

Terracing is the recommended treatment for sloping agricultural land, where erosion is most damaging. Stone or soil bunds are constructed to form bench terraces. Grassed waterways with check-dams are constructed to lead away surplus rain water. Studies have established that terraces are instrumental in reducing the rate of soil erosion.[15] Another effect of terracing is that the water retention capacity of the soil improves. Springs, which have been dried up for years, gradually return.

Closing of degraded hillsides is also practised. MOA and the local peasant association agree to close a certain hillside from further grazing, cultivation and fuelwood collecting. Peasant guardsmen enforce the closing. Sometimes trees and grasses are planted, but often the closure is left for natural recolonialization. The result is generally impressive. Vegetation returns within a few years. Cut-and-carry grass systems are introduced to harvest some of the vegetation.

Closing is instrumental in reducing the erosion from hillsides. A positive by-effect is reduction in gully formation. In existing gullies check dams built by stones and planting of grass and bushes prevent the gullies from further growth.

Planting of trees is also a prominent activity. There are some 50 MOA tree nurseries in the region capable of raising 35 million seedlings annually. In addition, there are more than a hundred small community-based nurseries with an estimated total capacity of 25 million seedlings per annum. Most of the seedlings raised in nurseries are actually

---

* The administrative classification refers to the one prevailing up to September 1987. It included 14 regions, 105 districts (awrajas) and some 500 sub-districts (woredas). With the adoption of the new constitution in late 1987 a new administrative system is being set up.

185

planted each year. Eucalyptus are most common followed by Cypressus and Acacia species. Some nurseries also multiply grass seeds for terrace stabilization and pasture improvement.

A number of problems beset these activities.

Terracing encroaches on the area available for cultivation and provides favourite hiding places for rodents.

In closed hillsides, species with low palatability and nutritional values tend to dominate. Another problem is that the grazing pressure is transferred to nearby hillsides which suffer doubly.[16]

Survival rates of tree seedlings is reportedly low, around 40%. The reasons are that planted species and provenances do not always match the local eco-type (altitude, temperature, soil depth, etc.) and that management (i.e. thinning) is deficient once plantations have been established.[17]

In 1987 MOA commissioned an evaluation of the community forestry programme in Ethiopia. Among the socio-economic reasons for the rather weak performance, the evaluation team found that peasants were reluctant to establish private woodlots due to fear of confiscation and due to uncertainty whether they (the peasants) would lose control of their holdings due to resettlement, villagization or formation of collective farms.[18]

The technical and biological aspects of these problems are given attention by MOA. Trials are conducted to find solutions.

In the early 1980s when environmental rehabilitation went to scale in Wollo and Gondar, the emphasis was on stone and earth structures. Gradually it has shifted to include vegetative measures. The next step, as envisaged in MOA planning documents, is to integrate crop and livestock production with conservation. When such activities have been tested and have shown productive results, conservation will be supplemented by production. Only then can one expect the peasants to become genuinely interested.

Terracing and tree planting are organized in large-scale food-for-work campaigns. Through the peasant associations people are mobilized to plant trees, construct terraces, etc. Technical supervision is provided by the Ministry of Agriculture, while The World Food Programme provides food, bilateral government agencies and NGOs provide technical assistance, hand tools, nursery supplies and transport.

Hundreds of millions of trees have been planted and tens of thousands of terrace-kilometres have been built. The motivating force is the food. Participating peasants are entitled to a daily ration of wheat (2–3 kg) and edible oil (120 grams). The food is delivered each month. The peasant associations and MOA are in charge of distribution which takes place monthly. Despite logistical problems with deliveries the system works reasonably well. No negative impact on local food production has been observed. The conservation work takes place during agricultural slack seasons—except planting of tree seedlings which must be done at the onset of rains. Due to the serious food shortages during the 1980s these deliveries have been a welcome supplement to the locally produced food.[19]

Donors are encouraged to participate in rehabilitation programmes. A number of donors, mainly NGOs, are active in Wollo. Amongst them are Red Cross societies working through the Ethiopian Red Cross Society and Christian organizations working through the Ethiopian Evangelical Church Mekane Yesus. SIDA is the major bilateral governmental agency and UNSO represents the UN family. The long-term presence of the voluntary organizations is uncertain since they rely on member contributions.

Food-for-work is the motivating force in people's participation. With support from the World Food Programme, bilateral donors and NGOs, MOA runs food-for-work activities in nine regions, involving some 800,000 persons and with a food subsidy of more than 100,000 metric tons per annum. It is the second biggest programme of its kind in the world.[20]

In the Wollo programme seven river catchment areas are included. They are situated along the main north–south trunk road and cover 7,200 sq.km. or 18% of the highland area in Wollo. Up to 1987 roughly 15% of this area had been 'treated' with some form of rehabilitation (trees, terraces or closures). The World Food Programme provides 20,000 metric tons of food to Wollo annually in support of the conservation activities.[21]

The activities mentioned above are impressive as such. But they affect only a small proportion of the highlands. Away from the major roads little conservation activities are to be seen. The great majority of highlanders live in densely populated but roadless areas. Two inter-related reasons for the lack of conservation activities—low official priority and peasant insecurity—will be discussed below.

Bias in development strategy

The Workers' Party of Ethiopia (WPE) systematically promotes a transition to socialism in the countryside. The major ingredients include state control of grain trade, establishment of Producers' Cooperatives and villagization. The Agricultural Marketing Corporation is the official procurement agency for grains. Prices are fixed and all peasant households have to deliver a quota. The Producer's Cooperatives are agricultural production collectives. Individual peasant households are encouraged to pool their resources collectively. The traditional dispersed settlement pattern is changed by moving all households of a peasant association into nucleated villages.

The achievements have been substantial in grain trade, where private traders either have closed business or function as agents for the AMC. Villagization has also proceeded far. More than eight million people now live in villages. Collectivization, however, is a disappointment to the officials. Despite continuous agitation, less than 3% of peasant households have joined Producers' Cooperatives.[22]

The aims of these policies are to create conditions for agricultural modernization and socialist transformation. The policies are pursued vigorously in the central and southern regions, where the agricultural potential is considered high.

The eroded highlands in the north are not considered to have an agricultural potential. Government policy has emphasized resettlement. This would, in the official thinking, serve the dual purpose of reducing land pressure in the highlands and mobilizing labour to exploit the vast lowland areas in the south and south–west, which traditionally have been sparsely populated.

When drought struck in 1984, the government carried out a large-scale resettlement campaign. Of the more than 600,000 persons resettled in 1984–86 some 380,000 were inhabitants of Wollo. Since 1987 resettlement has lost momentum. In 1987/88 only 3,500 people had been resettled. The migrants were composed of young married couples.[23] The pace of resettlement has slowed down partly due to lack of resources to administer the efforts and partly due to the realization among the policy-makers that resettlement creates many problems while it is not yet proven that it solves any.[24]

Resettlement is inevitable in Ethiopia. The question is where people should move and under which conditions. It has been argued that relocation within individual peasant associations would be the first step. The potential of the lowlands and their health-hazards are yet to be systematically explored.*

---

* Unfortunately, the attraction of resettlement from the peasant's point of view was considerably reduced by the rough methods employed by the authorities in 1984/85. Drought-stricken districts got quotas to fulfil. The local administration and army units made sure the quotas were achieved. In the process, people were rounded up in market places, children were separated from their parents and food aid was sometimes withheld until the local quotas of settlers had been fulfilled. The attraction of the programme did not increase when people arrived in settlement schemes, only to find that reception facilities were deficient, diseases were rampant and that they were supposed to become agricultural labourers rather than landowning peasants. Gradually, the adminisration of the schemes ahs improved. but it remains to investigate whether the costs and the administrative energy would have been more wisely spent in rehabilitation programmes in the north.

Since the government's development funds are consumed by agricultural programmes in the high-potential regions and by resettlement, not much remains for environmental rehabilitation in the north. As mentioned above, the capital costs of soil conservation activities are mainly funded by external donors.

Lack of commitment

The lack of local committment to environmental rehabilitation is indicated by the common sight of stone terraces which have collapsed and are not rebuilt.

One aspect of the inadequate committment is that the peasants are not fully convinced that conservation measures increase production. Terraces steal land from cultivation and attract rodents which feast on the meagre crops. Closed hillsides reduce grazing areas. The planted trees may not be touched.[25] Yet the peasantry is well aware of the fact that their cultivation methods cause erosion and they know that erosion is the cause of declining crop yields.[26]

The other aspect is insecurity. This problem relates to the overall development strategy pursued by the government. The official stress on socialist transformation has been mentioned above. Villagization, collectivization, grain quotas, etc. are designed and implemented in the conventional top–down fashion. There is little room for adaptation to individual and local conditions.

The peasants have little room for manœuvre. Despite the celebrated land reform in Ethiopia, peasants do not control the land. The reform abolished feudal obligations and gave use rights to former owners and tenants alike. But ownership is vested in the state. And the state has shown that it does not respect peasants' use rights when designing its grand schemes. No peasant in Wollo or elsewhere can know for sure whether the household he is part of will remain intact in the near future. The members of the household may be instructed to move into a village within the local peasant association or to a resettlement far away or to join a Producers' Cooperative. Although none of these alternatives may be worse than the present condition, the point is that the peasants cannot decide on their own. Therefore, the willingness to invest hard labour in environmental rehabilitation, unless it is supported by food-for-work, can only be negligible.

It can be argued that the peasants do not consider government sponsored environmental programmes a viable option. The missing link is development. Conservation of nature does not bring an improved living standard to peasants. If the programmes discussed above would expand in scale drastically, the peasants would be gravely threatened. As long as only a minority of the hillsides are closed, peasants can cope with it. But if closures expand to the dimension they would need in order to have a real impact on the environment, then the peasants' livestock would starve. Without oxen, there can be no tilling of land. Without goats, the last cash-convertible reserve would disappear. Such losses could not be compensated for by food-for-work.

Thus, the Ethiopian soil and water conservation activities, which constitute the most comprehensive large-scale environmental rehabilitation programme in Africa, can only accomplish bits and pieces. The dominant trend is still accelerated degradation.

POLITICAL CONSTRAINTS

The limited scope of environmental rehabilitation programmes also have a political dimension which is related to the unsettled political questions in the north. Although Ethiopia has a long history and political tradition, the attempts to build a unitary state are recent.

Problems of state building in Ethiopia

Until the late 19th century 'Abyssinia' was the name of a territory roughly corresponding to the mountainous parts of the present-day administrative regions Eritrea, Tigray, Gondar, Wollo, Gojam and the part of Shoa north of Addis Ababa. Within this territory, power was wielded by aristocratic warlords holding titles such as *negus* (king), *ras* (duke), etc. The social structure in these highlands had clearly "feudal" traits.[27] At the bottom were peasants obliged to render personal services and agricultural produce to their superiors and follow them in war. Political and military alliances shifted constantly and warfare was common.

A strong unifying factor was Monophysite Christianity which provided a cultural identity distinct from the muslims and pagans surrounding the highlands. Two major languages were spoken. Tigrinya in the north and Amharic in the south. Both are semitic languages and trace their origin to the liturgical language, Geez.

At the times of strong rulers, central administration expanded and the power of local lords was restricted. At other times real power was in the hands of the regionally based warlords.

In the early 19th century central political power was at a low ebb. There were kings in Tigray, which at that time included the southern highlands of present-day Eritrea, in Gondar, Gojam, Wollo and Shoa. In the latter half of the century centralized political control grew, partly as a response to European imperialism. Eventually Menilek, the king of Shoa, was crowned emperor and embarked on a series of combined diplomatic and military campaigns which gave him hegemony over Abyssinia and military control over vast tracts of land to the south of Addis Ababa. In this way the state of Ethiopia was created around the turn of the century. To the east, west and south the borders were vague. To the north Ethiopia clashed with Italian colonialism. Italy carved out the coastal strip and the Tigray highlands down to the river Mareb. This became Eritrea.

The latter part of Menelik's and all of Haile Selassie's reign can be characterized as vigorous attempts at state penetration. The boundaries were defined and internationally recognized. The name Ethiopia was given to the state. The power of regional lords was gradually diminished and centrally appointed governors were put in place. A European-type administration was introduced and feudal obligations were transformed into direct and indirect taxes. A thin network of physical infrastructure was built.[28] After the brief period of Italian occupation, the Ethiopian state has continued its programme of administrative penetration in the countryside. The pace was increased and the scope broadened after the revolution in 1974.

Efforts by the Ethiopian state to penetrate the territory and control the inhabitants do not proceed undisturbed. The most serious challenge to the Ethiopian state today is the wars in Eritrea and Tigray. Ironically, the war in Eritrea is closely related to the most conspicuous achievement of the Ethiopian state during Haile Selassie's rule. The former Italian colony of Eritrea was federated with Ethiopia in 1952 through a UN resolution. Eritrea was to have its own institutions for internal affairs. In actual practise, the Imperial government imposed its decisions on the federated territory with increasing force and in 1962 the Eritrean parliament was dissolved. Eritrea was incorporated into the Ethiopian state as an administrative region ruled by a governor appointed by Addis Ababa. From the point of view of the Ethiopian state an achievement was scored in terms of direct control over an expanded territory. This very fact created resentment and armed insurrection increased during the 1960s. During the Ethiopian revolution it was made clear that the military government was not prepared to give Eritrea a special status. The conflict then escalated. On the Eritrean side the Eritrean People's Liberation Front (EPLF) emerged as the major spokesman for independence out of a number of insurgent groups.[29]

During the last decade EPLF has built up a military apparatus which challenges the Ethiopian army in Eritrea. A civil administration is also operating in areas outside government control. EPLF claims the right to self-determination for the Eritrean people, including secession from the Ethiopian state.

During the last decade war has torn Eritrea. Almost every year the Ethiopian armed forces mount offensives ravaging the countryside. Counter-offensives are then pursued by the EPLF army and government troops are rolled back.

No less serious for the Ethiopian government is the challenge from the Tigray People's Liberation Front (TPLF). The region of Tigray has traditionally guarded its autonomous status and resisted government by Addis Ababa. When the military government dismissed the traditional aristocratic lord as regional governor in 1975 and showed signs of extending direct control, opposition gained momentum. TPLF emerged as the major resistance movement in the late 1970s and is now a military factor of importance. The Ethiopian army and administration hold only some towns and trunk roads. Military operations roll back and forth devastating the countryside. Most parts of the region lack effective administration while TPLF attempts to impose its own order. TPLF is not for secession. It considers Tigray to be an integral part of Ethiopia. But it considers the present government in Addis Ababa to be illegitimate.

Implications for environmental rehabilitation activities

Political unrest makes environmental rehabilitation difficult or impossible. Tigray is worst affected. The mountain slopes, which have been grazed and cultivated for millenia, are worn out. Degradation in Tigray is on the verge of having irreversible consequences, i.e. agricultural land may be lost for good. If any region would need environmental first aid, that would be Tigray. It should be noted though, that there are soil and water conservation projects along bits and pieces of the main road. These are part of the World Food Programme activities. The activities focus on water conservation (dams, ponds).

Due to the warlike situation, no systematic rehabilitation work can be mounted. While the Ethiopian administration cannot reach the rural areas, the TPLF has very few resources for non-military activities.

A related discussion holds for Eritrea. Here, the government controlled areas are more extensive than in Tigray. MOA operates soil and water conservation activities along the major roads.

One encouraging observation is that neither TPLF nor EPLF appear to harass the government technicians working with environmental rehabilitation. They realize that these activities are necessary. When government staff get caught by guerilla units, they are usually released immediately and encouraged to carry on their work. On the other hand, the opposition movements are ambivalent to food-for-work. It is argued that food deliveries by the Ministry of Agriculture tend to foster a more favourable attitude to the government.

The regions immediately south of Tigray are Gondar and Wollo. Here the situation is mixed. North of the "Chinese" road cutting east-west through both regions, insurgency is common. TPLF operates here and so do other opposition groups. Hit-and-run operations are directed at military and administrative centres. Effective Ethiopian administration is therefore reduced. South of the road only sporadic incidents have been reported. The Ethiopian administration is unchallenged here.

Government support to environmental rehabilitation in Wollo is contradicted by the general agricultural policy (compulsory procurement of quota grains at low prices, resettlement, villagization, collectivization) which appears to exploit the meagre resources of Wollo rather than to replenish them. Even though there is an appreciation of the need for rehabilitation in the highest political circles, the government's heavy-handed way of implementing agricultural and environmental programmes is bound to create frustration. People are seldom consulted in a genuine way. They are told what to do and opposition is considered a counter-revolutionary attitude. Consequently, peasants participate in enviromental rehabilitation when food-for-work is arranged. Their relation to the rehabilitation programme is that of paid workers (in Amharic the schemes are called "Work for Food"), not of responsible landowners.

CONCLUDING DISCUSSION

The technical and biological aspects of soil and water conservation are known and MOA activities are gradually oriented in this direction. With a determined research effort, rehabilitation/production packages could be designed for different agro-ecological zones. The tripartite administrative set-up consisting of peasant associations, MOA and donors has been successfully tried in Wollo and elsewhere.

The crux of the matter is thus political rather than administrative or bio-technical.

In order to achieve durable results in the combat against environmental degradation, the Ethiopian state would need to marshal substantial economic and human resources as well as popular support. At present the resources are not available because priority is given to army and security.

The government diverts the lion's share of state resources to combat the insurgents and to reassert its authority.[30] As a result, the official development budget is squeezed. Moreover, the development programmes take on increasingly *extractive* (procurement of grain at low prices, collection of taxes, unpaid labour campaigns, financial contributions to 'revolutionary' tasks) and *penetrative* (relocation of settlement through villagization and resettlement, collectivization of agriculture, establishment of state-controlled mass organizations) aspects. This is because the overriding aim is to control the population and agricultural production in order to finance the wars.

Western donors have always watched the revolutionary government with restraint. Ethiopia gets less development aid per capita than any other third world country. Only when fully-fledged famines ravage the country is generous but short-term assistance forthcoming. From a technical and administrative point of view, Ethiopia has a good reputation as recipient of aid. The absorption capacity is considered to be one of the highest in Africa. The donors' hesitation is thus clearly political. It would not be far-fetched to assume that substantial funds for development and environmental rehabilitation could be mobilized, if they were accompanied by less dogmatic Marxist–Leninist policy and practise.

It can be argued that a precondition for large-scale environmental rehabilitation in the northern highlands is a political settlement between the government and the opposition movements. First, it would make vast areas accessible for rehabilitation programmes. Second, it would free resources for rehabilitation and development. Third, it would ease the officially felt need to demand tributes from the peasantry under all circumstances and create preconditions for more producer-oriented policies. Fourth, it would attract increased assistance from donors.

NOTES

1. Ministry of Agriculture and FAO (MOA-FAO) 1984: *Ethiopian Highlands Reclamation Study. Summary and Working Papers*. Addis Ababa.
2. Simoons, F.J 1970: "Some Questions on the Economic Prehistory of Ethiopia", in Fage, J.D. and Oliver, R.A., *Papers in African Prehistory*. London.
3. Among the many scholarly works on early Ethiopian history including the Axumite, Zagwe and Abyssinian kingdoms, I would recommend Haberland, E. 1965: *Untersuchunqen zum Äthiopischen Köniotum*. Wiesbaden. Trimingham, J.S. 1952: *Islam in Ethiopia*. London, and Bartnicki, A. & Mantel-Niecko, J. 1978: *Geschichte Äthiopiens Vols 1 and 2*. Akademie-Verlag, Berlin.
4. Taddesse Tamrat, 1966: *Church and State in Ethiopia 1270- 1527: Pankhurst, R. Economic History of Ethiopia 1800–1935*. Addis Ababa.
5. While the state provided the military and political framework for expansion, agricultural technology, management and property systems were introduced and reproduced by the Christian peasant communities. A stimulating anthropological

approach to the analysis of farming systems in northern Ethiopia is provided in McCann, J. "History, drought and reproduction: dynamics of society and ecology in northeast Ethiopia", in Anderson, D.M. and Johnson, D.H. 1978: *The Ecology of Survival: Case Studies from North African History*. Lester Crook Academic Publishing.

6. Central Statistical Office, 1985: *National Census*. Addis Ababa.
7. Personal communication from Anders Tivell. 8. MOA-FAO, *op.cit.*
9. Adams, M. 1987: "Primary Health Care and Family Planning in Conservation-Based Development", *Working Paper No. 2 : SIDA Support to Welo Region—Background Papers*.
10. Ethiopian Red Cross—Upper Mille and Cheleka Catchment Disaster Prevention Programme, 1988: *Rapid Rural Appraisal—A Closer Look at Life in Wollo*.
11. Agricultural Marketing Corporation, 1986: *A Brief Note on Grain Marketing Policy*. Addis Ababa. Hultin, J. "The Predicament of Peasants in Conservation-Based Development" *Working Paper No. 8; SIDA Support to Welo Region, op. cit.*
12. Dessalegn, Rahmato, 1987: *Case Study from North–east Ethiopia*. Food and Famine Monograph Series No. 1. Institute of Development Research, Addis Ababa University.
13. *Ibid.*
14. Information from MOA.
15. The University of Bern runs a collaborative programme with MOA investigating the effects of different physical and vegetational conservation measures on sloping agricultural land.
16. Fruhling, P. 1988: *Utveckling bättre än nödhjälp. Om Röda korsets katastrofföre-byggande arbete i Wollo*; Hultin, J. 1988: *Farmers' Participation in the Wollo Programme*.
17. SIDA Soil Conservation/Community Forestry Welo Mission, *SIDA Support to Welo Region*. Background Papers Nos. 1 and 3 1987.
18. Markos Ezra and Kassahun Berhanu, 1988: *A Review of the Community Forestry Programme and an Evaluation of its Achievements*. Addis Ababa.
19. Office of the National Committee for Central Planning, 1986: *Workshop on Food-for-Work in Ethiopia*. Addis Ababa.
20. *Ibid.*
21. SIDA SC/CF Welo Mission, *Proposed SIDA Support, Annex 3*.
22. I have analyzed the agricultural policies in the high potential areas at length elsewhere. See Ståhl, M. "Capturing the Peasants through Cooperatives. The Case of Ethiopia", in *Northeast African Studies* (forthcoming).
23. Hultin, *op. cit.*
24. Hultin, 1988, *op. cit.*
25. Fruhling, *op. cit.* Hultin, 1988, *op. cit.*
26. Yeraswerk Admassie and Solomon Gebre, 1985: *Work in Ethiopia. A Socio-Economic Survey*. IDR Research Report No. 24 Addis Ababa. Fruhling, *op. cit.*
27. Teshome Kebede, 1984: "Some Aspects of Feudalism in Ethiopia", in Rubenson (ed), *Proceedings of the Seventh International Conference of Ethiopian Studies*. Scandinavian Institute of African Studies. Pankhurst, R. 1966: *State and Land in Ethiopian History*. Oxford University Press.
28. Clapham, C. 1969: *Haile Selassie's Government* Longmans. Perham, M, 1968: *The Government of Ethiopia*. Faber and Faber.
29. For an analytical account of the emergence of liberation fronts in Eritrea, see Markakis, J. 1988: "The Nationalist Revolution in Eritrea", *The Journal of Modern African Studies*. 26.1.

30. Although no official figures are available, it is estimated that 40–50% of government spendings are consumed by the military apparatus and another 10% by security.

Population and Land Degradation

A Political Ecology Approach

Christer Krokfors

INTRODUCTION

Within the last decade there has been a marked growth of interest in the quality of African environments, the disruption of the natural ecosystems and the depletion of natural resources. Different aspects of land degradation and its causes in African societies have been projected from the cloistered world of science into the forefront of international debate. All aspects of the use of Africans of their environment have been widely discussed with great interest. The speed and nature of environmental change (both natural and man-induced changes) in recent years have brought about a series of environmental problems of continental magnitude—including population growth, energy scarcity, the provision of food supplies and clean water, new forms of land exploitation and climatic changes being the most commonly discussed.

A lexicon of general crisis has entered popular and professional discourse almost without critique. This is still somehow the state of the art in the latter part of the eighties. As a result of this lexicon there has been a development of a profound contradiction between a deterministic (objectivist) and a voluntaristic (subjectivist) ways of interpreting the conditions of the African societies. The two interpretations refer to different scientific paradigms. The first typically falls within the purview of natural sciences and tends to over-emphasize technological solutions to the crises. The second again lies within the social sciences and approaches developmentalist solutions. There are considerable epistemological and ideological problems in overbridging the two. This means also that information on African environments and analyses of environmental processes are extremely shattered and confusing.

The approach to African population and environment used in this paper is a political ecology one, and the aim is to provide a conceptual framework for an understanding of the relations between population and land degradation. I like to stress the word *political* because the focus will be on the politicized relations in which population and environmental resources are embedded, thus grounding ecology in the web of social relations (cf. Watts, 1985). Political ecology as a concept is not common in the contemporary debate. It was even heavily criticized during the seventies and those who used the concept were said to be insisting on totality in explaining for instance the relations between population growth and global crises. Political ecology was accused of being a "methodological jumble" and leading to "quick technological fixes" (cf. Enzensberger, 1973). Others, like Hoffmeyer (1975), saw political ecology as a means of capitalistic exploitation and as a foundation for reductionistic biologism.

POLITICALIZATION OF POPULATION AND ENVIRONMENT

Since the seventies our understanding of African societies and their relations to environmental resources has deepened and led to a recognition that these relations are highly politicized. The way the concept "political ecology" is used in this paper should be construed against that background. The approach thus encompasses the interactive effects of population–environment, and contributions of different geographical scales and hierarchies of socio-economic organizations (cf. Blaikie, 1985). However, the emphasis will be on the local level processes in specific locations and on specific segments of population. The overall framework is a discussion of land degradation in relation to survival opportunities. In this case it is important that population is interpreted not only in the demographic sense as numbers and rates—rather, population should be seen as consisting of politicized social structures. The physical environment is relationalistic to these structures, i e population structure and differing use of environmental resources are two sides of the same coin. These differing interactions between population structure and environment create either accumulation of wealth or a tottering ground for survival. That means that also the physical environment is politicized.

A first approximation of the politicized population/environment situation is illustrated in Figure I. No geographical scale is yet assumed but the figure is best interpreted as representing a local situation. The figure shows a population divided along a poverty continuum into two groups, one rich, one poor. Likewise there is an environment divided along a resource quality continuum into a quality-rich and quality-poor section. The resource access field shows that the rich minority of the population for one reason or another has access to the quality-rich environment while the poor majority of the population only has access to the very quality-poor environment.

This simple figure shows that it is through the resource access field that the processes of politicization of both population and environment are taking place, that is who is able appropriate surplus value, and who can only achieve a meagre existence. The difference in access to land and land use between these two groups will be the background to the discussion about population and land degradation. However, this requires an analysis of the interacting qualities of both the environment and the population.

LAND USE AS A RESOURCE

Land as an environmental resource is a complex phenomenon. Environmental (physical) resources are a normally described either as stock resources or flow resources (Rees, 1985; Saetra, 1975). Stock resources are either abiotic or biotic. Their common characteristic is that once used they will not be renewed within a foreseeable timespan. Flow resources are if their base of production remains intact renewed within a sufficiently short timespan to be of relevance to human beings.

Land is a flow resource continuously produced by abiotic and biotic processes. The value of land is best expressed through its capability, "the intrinsic qualities to satisfy a particular use" (Blaikie and Brookfield, 1987). As long as these qualities are reproduced at the same pace as they are used there is no land degradation. However, in specific socio-economic situations land *has to be* over-used. The natural processes are not able to reproduce, for instance, valuable plant nutrients at the same pace as they are extracted. In this case, land has changed from a flow resource to a stock resource, the intrinsic qualities of which are soon extracted and what is left is degradaded land.

Figure I shows that different sections of the population have different access to, in this case, land of differing quality. High quality land poses a set of access qualifications that can only be met by a part of the population, for instance through capital or technological assets, membership of some influential group, inheritance, etc. The return of this land is high, its capability usually well maintained and there is a low risk of land degradation.

Low quality land has low access qualifications (marginal areas, waste land, slopes, peri-urban land and the like). At the same time this land is the most asked for (quantitatively), has a low return and is therefore easily over-used and degraded. These differences in access to land are characteristic of most of Africa. The part of the population that has access to high quality land is often also that part that has benefited from various types of development projects like introduction of high yielding varieties, irrigation, well development, cash cropping, etc. This has often also meant that other parts of the population are forced to accept low quality lands, or have access only to landholdings that have a size which is to small to warrant the household's subsistence. In such a situation the risk for land degradation increases enormously and even new strategies for survival have to be developed.

LAND DEGRADATION AND SURVIVAL

Africa is often described as the continent of peasants. Several studies on peasants' relations to environmental resources have shown a great variety of adaptive strategies and means of coping with environmental stresses (cf. Richards, 1983, 1985, 1986; for pastoralists Dahl and Hjort, 1976; Dyson–Hudson, 1980). In a time of "African crises" one gets the impression that these adaptive strategies and mechanisms have disappeared. To get an understanding of the contemporary situation one has to look into what it means to be a "peasant" under changing circumstances.

The literature on peasant and peasant societies is voluminous and contradictory (cf. Alpers, 1985; Bradby, 1975; Foster–Carter, 1978; Kitching, 1985), there is not even an agreement on the concept "peasant". For the purpose of this paper peasants or peasant household are characterized as those producing mainly for the reproduction of the household, that is use value. However, there are hardly any pure peasants in Africa nowadays (with the possible exception of some secluded groups—see below). Since colonial time, or even earlier, new modes of production, mainly different forms of capitalism, have entered Africa, and exist along with or superimposed on earlier pre-capitalistic modes. From the political ecology point of view this means that the environmental resources and especially land are under new forms of pressure. The capitalistic mode of production aims at accumulation of wealth and is mainly based on inputs generated through extraction of stock resources. The pre-capitalistic mode in its pure form was based on harvesting flow resources. The articulation of these two modes means that both the productive activities in the altered pre-capitalistic mode and its relation to land is in a contradictory stage. This is manifested through a complex system of interactions between people, people and land, and between different geographical areas.

Figure II is a simplified graphic representation of articulation of a dominated peasant or pre-capitalistic mode to a dominating capitalistic mode of production. The geographical scale is local in the sense that there are interactions both towards a national/international level, and within the same level urban–rural continuum. The articulation of the two modes of production creates a double demand on the environmental resources: on the one hand supply of use value to the peasant segment, on the other hand an extension and intensification of commodity production in the emerging capitalist segment. Thus a market for the commodities has to be created through increases of productivity of land and labour in the peasant segment and some of the survival elements have to be appropriated through exchange relations.

From a political ecology point of view this double demand implies that land use becomes contradictory in the sense that the low development of the productive forces within the peasant agriculture has to compete over access to land with the higher developed forces of the capitalist more intensive form of land use techniques. In a situation where capital has a general interest in extension and intensification of commodity relations there is usually a tendency of "opening up" the peasant controlled land through an introduction of rural development schemes aimed at increasing the productivity of peasant labour or replacing it with capitalist agricultural or livestock enterprises. It is in such situations the traditional adaptive strategies of the peasants cease to function. This situation is also expressed through a relational organization of space that correspond to the articulated modes of production—competition between different socio-economic groups is expressed in differing competitive uses of space. Figure II can also serve as a background to a discussion about three prominent features related to population and land in contemporary Africa: (i) the prospects of peasants, (ii) the geographical transfer of value, and (iii) development of new survival strategies.

The prospects of peasants

Bernstein (1979) has dealt with the situation of African peasants in their role as rural producers securing their livelihood through the use of family labour on family land. However, the peasants are not discussed in isolation. Bernstein clearly points out that peasants in Africa today are exploited through relations of commodity production and exchange which lock them into the international capitalist economy. It is against this background that the relation between peasants and land degradation has to be construed. Bernstein identifies what he calls "the simple reproduction squeeze" among peasant households. The "squeeze" is the effects of commodity relations on the economy of the peasant households "that can be summarized in terms of increasing costs of production/decreasing returns to labour". Production of items for exchange has become to a certain extent an integral part of peasant survival. Deterioration of terms of trade between commodities produced for the market and items of necessary domestic consumption acquired from the market is transmitted to the household economy in terms of reduction in consumption or an intensification of commodity production (i.e. through intensification of land use), or both. Because land and labour are the main means of production, the "squeeze" implies exhaustion of both. "The low level of development of the productive forces in peasant agriculture means that the households are extremely vulnerable to failure in any of its material elements of production". Changes in prices of the produce can then result in a sort of super-exploitation of land that destroys the capability of the land to satisfy the needs of the peasant household. A situation of sub-subsistence develops and the members of the household have at least partly to find new, often not land-based activities for survival. This means new social relations of surplus extraction, less attention is paid to the maintenance of land and continuous loss of land capability to satisfy what so ever need.

The mechanisms described above act to intensify the labour of the peasant household to maintain or increase the supply of commodities without costs for management or supervision (Bernstein, op.cit.). This means an appropriation of value by the dominating mode of production (see Figure II). Some authors (cf. Sender and Smith, 1986) interpret this as a transition to capitalism and the existing contradictions and crises in Africa as merely growing pains in this transition.

The geographical transfer of value

Land degradation in relation to the socio-economic structure of the population is best studied in a local setting. But this setting has to be related to higher levels of geographical and socio-economic organization. Most countries show an uneven economic development. This is mainly due to the increased use of inputs from stock resources by the dominating capitalist mode of production. On a regional level extraction of stock resources tends towards concentration of economic activities—the extracted products have to be brought together to form utilities. This causes an uneven distribution of capital accumulation based on transfers of values appropriated between regions of a country.

This spatial differentiation in accumulation can be attributed to two factors. One is the survival of pre-capitalist modes of production due to uneven spread of capitalist mode of production. The other is the qualitative differences within the capitalist mode due to different development of the productive forces and thus the ability to compete (Forbes and Rimmer, eds., 1984). Again there will be a contradiction between harvesting flow resources and extracting stock resources. The higher the development of a region's productive forces, the more important are the inputs from stock resources, and the greater the appropriation of value from other regions. The values transferred are rent, interest and profits derived from two basic sources: the extraction of surplus produce from the pre-capitalist mode of production and the extraction of surplus labour form the production processes. Thus the geographical transfer of values reinforces exploitation of land and labour at the local level; some areas being left with the only possibility of harvesting flow resources with a constantly decreasing return. In the long run, the only resource left for these areas are labour that has to be exported to areas of capital accumulation, the only mean of survival for the household.

Development of new survival strategies

Inequality in access to land and assets is the main reason for land degradation. Once the degradation process reaches a level where the land capability cannot meet with the demands from the land user he has to find other means for survival than land based production. As has been shown this is often the case when land use practices mainly based on harvesting flow resources confront with those using inputs based on extraction of stock resources. The result is a sub-subsistence situation. To manage such a situation the members of a household have to find new strategies of survival. These can either be adaptation to the new socio-economic situation like the "multi-active household strategy", or disclaiming from the contemporary situation like the "secluded-group strategy".

The **multi-active household** exists in rural as well as in urban areas. The members of the household engage in different income creating activities. Some might be land-based, others wage-based but very often the activities are of spontaneous and informal nature. Commonly the household is not only multi-active but also multi-localized. That means that the members earn income in different localities often of several days time-distance from each other Very often the household members use the different income opportunities which are existing within the rural–urban continuum (Hjort, 1979; Hesselberg, 1985; Bjeren, 1985). The interlacing of activities within the household enables the household to survive as an economic unity. In such a situation a large household with several income creating members is an advantage. It is therefore not surprising that those countries in sub-Saharan Africa that have the highest development of the productive forces in the capitalist sector also have the highest rates of population increase (Kenya, Zimbabwe, Ivory Coast). In these countries the demand for commodities from the market economy is highest and the simple reproduction squeeze most pronounced. The more members of the household, the more production that can be changed against commodities.

Through being multi-localized the household also has a multi-resource base. This is a significant aspect of the performances in Africa. The engagement of the household's members in different activities at different places using different resources is the core of the survival strategy (Krokfors, 1988). The members being in different places will be entitled to resources with differing access qualifications, their practical consciousness determining their resource use (Elwert, 1983). This implies new relationships between the people and the resource base, i.e. the physical environment. Resource uses that once were described as clearly belonging to a non-monetarized economy (i.e. "green leaves" in the diet) might suddenly be commercialized. The *one* member of a household in relation to members of other households has to use his innovative capacity to get access to resources that in the best way contribute to the survival of his own household. Locally this entails a totally new form of man–environment relations. It is not any more a question of multiple use of the environment by united households. Instead there is a competition for entitlement to resources in the same place by single-active members of multi-active, multi-localized households. In Africa this is a new man–environment relation, the ecological, demographic, and political consequences of which still are to be assessed.

The **secluded-group strategy** implies disclaiming from a society characterized by a dominating mode of production and the simple reproduction squeeze. Seclusion is based on a new societal ideology, a superstructural reconstruction aimed at ushering the local population into a radically new community and thus avoiding encapsulation in a wider capitalist system. This reconstruction is often based on religion: Christian, Islamic, traditional African, or some mixtures of these (van Binsbergen, 1981; Davidson, 1983; Lewis, 1955, 1956). The superstructural reconstruction is combined with an infrastructural reconstruction that entails new relations of the group's members to each other and to the means of production, i.e. land. This infrastructural reconstruction derives its strength through revitalizing the "economy of affection" based on reciprocal assistance, local control of the means of production and a concern with production of use value for the community before involving in exchange relations and taking risks (Hydén, 1983; Rudengren, 1981).

The secluded groups in Africa have a long history going back to the pre-colonial period and only a few aspects of political relevance for maintenance of land capability are dealt with in this paper. Seclusion attracted much attention in connection with the independence of the African states and in relation to the type of economic policy practised during the post-independence period. Through the emphasis on the local community and its survival under a specific superstructure different from that otherwise characterizing the post-independent state, the secluded groups often have made considerable economic progress in comparison with what is the case of groups under the simple reproduction squeeze. This also means that members of secluded groups pay much more attention to the maintenance of land capability than what usually is the case in the African states. Land has often been reclaimed in remote areas (Clarke, 1982; Davidson, 1983; Hisket, 1984 for cases in Islamic West Africa, Lewis *op. cit.* ; Seger, 1986 for Somalia) or illegally occupied (i e van Binsbergen *op.cit.*). By the state authorities the secluded groups are often looked upon as a to threat the country's unity and therefore counteracted (see for instance van Binsbergen, *op.cit.* about the Lumpa rising in Zambia, and Coulson (1982) about the Ruvuma Dvelopment Association, Tanzania, that developed an Ujamaa-ideology and production infrastructure too radical to be accepted by the state authorities).

The multi-active household strategy and the secluded group strategy are the extremes of survival strategies in contemporary Africa. The former is an adaptation to, the latter a reaction against, the simple reproduction squeeze. Thus, also the political consequences of the two extremes are very different, as are their abilities to maintain land capability and to provide for daily supply of utilities.

CONCLUSIONS

The main purpose of this paper has been to show that interlacing qualities of population and land determine the degree of land degradation. Figure III shows the relation of the political ecology approach used in this paper to the more common approach dealing with the demographic characteristics of population and their supposed land use effects. In the political ecology approach the politicized social structures and land capabilities are construed as those influencing the proximate determinants of both the demographic processes and the possible processes of land degradation. The qualities of land were grounded in the web of social relations and it has been stressed that affiliations to different socio-economic groups also determine if the land user has to over-exploit the land. In a situation when the land's capability cannot satisfy the needs of the user, new survival strategies have to be found. The ecological consequences of the new survival strategies are still to be assessed.

REFERENCES

Alpers, E. 1985: "Saving baby from the bath water". *Canadian Journal of African Studies, 19.* pp. 17–18.

Bernstein, H. 1979: "African peasantries: a theoretical framework". *The Journal of Peasant Studies,* 6:4, pp. 421–443.

van Binsbergen, W.M.J. 1981: "Religous change in Zambia. Exploratory studies"". London.

Bjeren, G. 1985: "Migration to Sashemene. Ethnicity, gender and occupation in urban Ethiopia". Uppsala.

Blaikie, P. 1985: *"The political economy of soil erosion in developing Countries""*. London.

Blaikie, P and H. Brookfield, 1987: *"Land degradation and society".* London.

Bradby, B. 1975: "The destruction of the natural economy". *Economy and Society,* 4, pp. 127–161.

Clarke, P.B. 1982: *"West Africa and Islam".* London.

Coulson, A. 1982: *"Tanzania. A political economy".* Oxford.

Dahl, G. and A. Hjort, 1976: "Having herds: Pastoral growth and household economy". *Stockholm Studies in Social Anthropology* .

Davidson, B 1983: *"Modern Africa".* London

Dyson–Hudson, H. 1980: "Strategies of resource exploitation among East African savanna pastoralists". In D. Harris (ed.) *Human ecology in savanna environments.* London, pp. 236–254.

Elwert, G. 1983. *"Bauern und Staat in Westafrika. Die Verflechtung sozio-ökonomischer Sektoren am Beispiel Benin".* Frankfurt.

von Enzensberger, H.M. 1973: "Zur Kritik der politischen Ökologie". *Kursbuch* 33, pp. 1–42.

Forbes, D.K. and P.J. Rimmer (eds.): "Uneven development and the geographical transfer of value". *Human Geography Monograph* 16, School of Pacific Studies, The Australian National University, Canberra.

Foster–Carter, A. 1978: "The modes of production controversy". *New Left Review* 107, pp. 47–77.

Hesselberg, J. 1985: "The Third World in transition. The case of the peasantry in Botswana". Uppsala.

Hiskett, M. 1984: *"The development of Islam in West Africa"..* London.

Hjort, A. 1979: "Savanna town. Rural ties and urban opportunities in Northern Kenya". *Stockholm Studies in Social Anthropology* .

Hoffmeyer, J. 1973: *"Dansen om guldkornet. En bog om biologi og Samfund.* Copenhagen.

Hydén, G. 1983: *"No shortcuts to progress. African development management in perspective".* London.

Kitching, G. 1985: "Suggestions for a fresh start on a exhausted debate". *Canadian Journal of African Studies,* 19:1, pp. 116–126.

Krokfors, C. 1988: "The unique versus the general—some lessons for development geography". *Occasional Papers of the Finnish Association for Development Geography,* 23, pp. 1–9.

Lewis, I. 1955: "Sufism in Somaliland. A study in tribal islam I". *Bulletin of School of Oriental and African Studies* 17:3, pp. 581–602. 1956: "Sufism in Somaliland. A study in tribal islam II".

Ibid. 18:1, pp. 145–160

Rees, J. 1985: *"Natural resources. Allocation, economics and policy"*. London.

Richards, P. 1983: "Ecological change and the politics of African land use". *African Studies Review*, 26, pp. 1–72, 1985: *"Indigenous agricultural revolution: ecology and food production in West Africa"*. London 1986: *"Coping with hunger: hazard and experiment in an African rice-farming system"*. London.

Rudengren, J. 1981: *"Peasants by preference? Socio-economic and environmental aspects of rural development in Tanzania"*. Stockholm.

Saetra, H. 1975: *"Den økopolitiska sosialismen"*. Oslo.

Seger, N. 1986: "Organisation of the irrigated agricultural economy amongst the small farmers of the Mubaaraak/Lower Shabelle region". In P. Conze & T. Labahn (eds.) *Somalia. Agriculture in the Winds of Change*. EPI Dokumentation Nr. 2, pp. 153–163.

Sender, J. and S. Smith, 1986: *"The development of capitalism in Africa"*. London.

Watts, M. 1985: "Social theory and environmental degradation". In Y. Gradus (ed.) *Desert development. Man and technology in sparselands"*. Dordrecht.

FIGURES

Figure I. *Politicalization of population and environment*

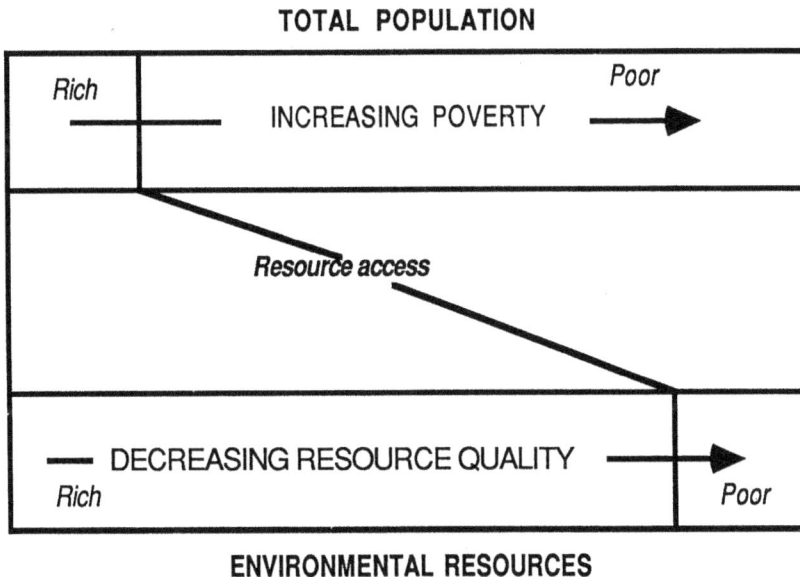

ENVIRONMENTAL RESOURCES

Figure II. *Creation of conflicting social structures and spaces*

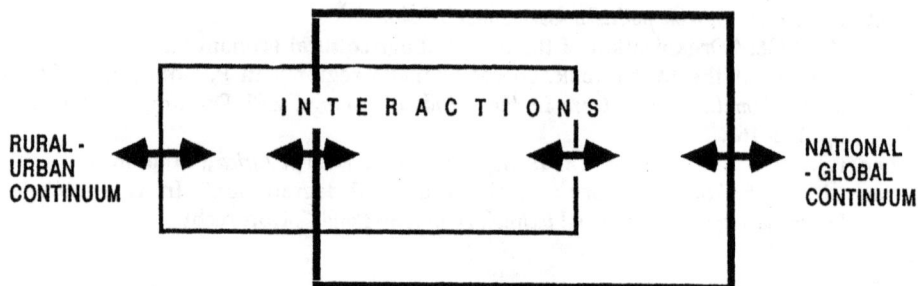

DOMINATED MODE	DOMINATING MODE
SIMPLE REPRODUC- TION SQUEEZE	APPROPRIATION OF VALUE
MULTI-ACTIVITY	SPECIALIZATION
FLOW RESOURCES TO STOCK RESOURCES	STOCK RESOURCES TO FLOW RESOURCES
CAPABILITY LOSSES	CAPABILITY MAINTAINED

Figure III. *The relationship between demographic characteristics and land development*

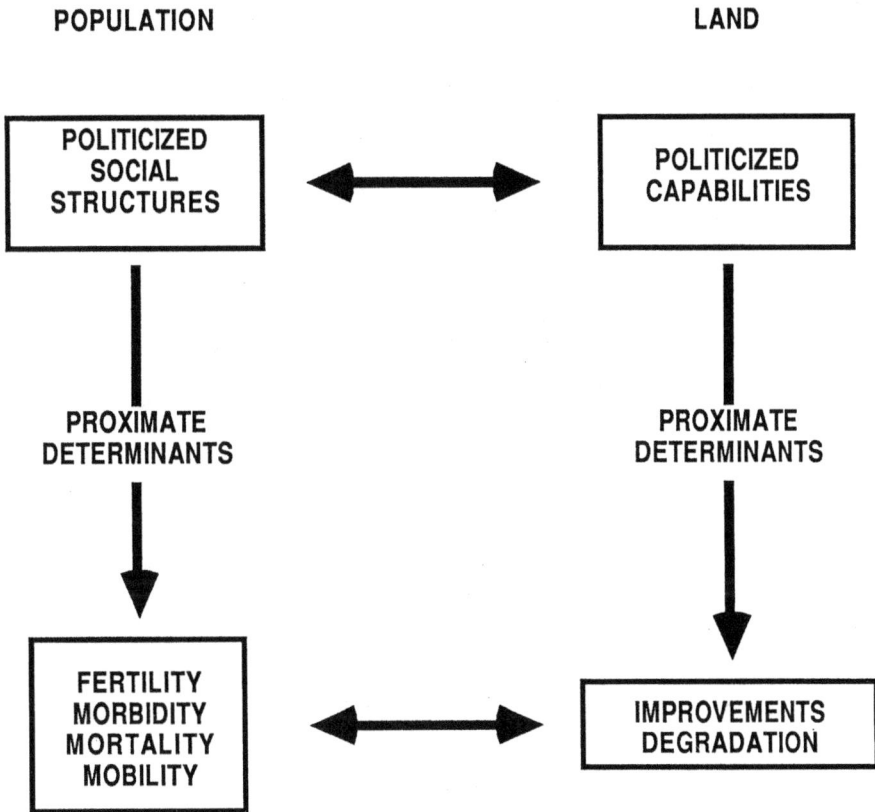

POPULATION

LAND

POLITICIZED
SOCIAL
STRUCTURES

POLITICIZED
CAPABILITIES

PROXIMATE
DETERMINANTS

PROXIMATE
DETERMINANTS

FERTILITY
MORBIDITY
MORTALITY
MOBILITY

IMPROVEMENTS
DEGRADATION

Population Growth, Environmental Decline and Security Issues in Sub-Saharan Africa

Norman Myers

"The picture for the period ahead is almost a nightmare. The potential population explosion will have tremendous repercussions on the region's physical resources such as land and essential social services. As a result of socioeconomic difficulties, riots, crimes and misery will be the order of the day. With weak and fragile socio-political systems, the sovereignty of African states will be at stake." (Dr. Adebayo Adedegi, Executive Secretary of the Economic Comission for Africa, 1983)

"Africa may be the major area of conflict in the 21st century. You have the possibility of real, serious catastrophe—in terms of human suffering, in terms of violent conflict and in terms of a retrogression in development. But these factors can be reversed if the world can collectively take care of resource distribution, population control, and control of the environment. Otherwise Africa will become a destabilizing factor in the equation of international peace and security." (General Olusegun Obasanjo, former Head of State of Nigeria, 1987.)

ABSTRACT

Sub-Saharan Africa is the one region of the developing world where population growth rates are still rising. Almost half of the increase in humanity's numbers during the foreseeable future is projected to occur in this region alone, which in turn is projected to feature a four-times increase in numbers. Yet there are signs that already the region suffers from population pressures, and to some preliminary extent these signs are borne out by analysis of alleged linkages between "excess" numbers and environmental decline. This arises especially and most clearly with respect to the natural-resource base that supports food production. But the linkages also relate to other manifestations of population problems, such as urban over-crowding, unemployment, pervasive poverty and the refugee phenomenon. Equally important, the linkages connect up with economic viability, political stability and ultimately with security concerns. The demonstrable interactions involve certain mediating mechanisms at work, in the form of social and institutional processes. In short, we can invoke a conceptual rationale, backed up by empirical evidence from a range of counties (e.g. Sudan, Ethiopia,

Kenya, Mozambique and Angola), to illuminate and clarify the putative capacity of population growth to serve as a destabilizing force when it operates in conjunction with environmental degradation.

INTRODUCTION

Can the ultra-rapid growth of population in sub-Saharan Africa (Table I) lead to conflict of various sorts? Is the region, with the highest population growth rates anywhere (Table II), thereby more likely to suffer political upheaval, civil disorder and even violent confrontation?

Moreover, if there is indeed a connection between population and conflict, how does it work? What is its "operational chemistry"? Do population problems directly and inevitably lead to violence? Or do they work indirectly, e.g. in catalytic conjunction with other factors such as environmental carrying capacity? If the latter, does the "other factor" complication make population itself less potent as a source of conflict? Or does it make it all the more dangerous, in that population pressures then work in less overt, hence less heeded, fashion?

Consider that almost 200 million people in sub-Saharan Africa are reckoned (World Bank, 1988) to receive less than 90 percent of the minimum of 2,200 calories a day needed to support an active working life—that is, they are chronically undernourished. Yet today's population of 508 million is projected to reach 678 million by the year 2000. This is an ultra-rapid rate of growth that operates in conjunction with adverse climatic conditions such as have obtained in much of the region for the past two decades. If these adverse conditions persist (they could well be aggravated by the climatic vicissitudes entrained by the greenhouse effect) (Farmer and Wigley, 1985; Glantz, 1987), the enfamished throngs that totalled 30 million in 1985 could well increase to twice as many by the early 1990s, and to a vast 130 million by the year 2000 (McNamara, 1985). This means that the proportion of starving people may expand from less than 7 percent of the region's population in 1985 to 19 percent by the year 2000. Moreover, if the region continues to sink into widespread depravation, will this not supply conditions for political turmoil within dozens of countries (Myers, 1987; Revelle, 1988)—and thus offer scope for general instability and threats to security, including Gadaffi type adventurism?

But while there are some ostensibly "obvious" linkages between population growth and violent conflict, this is an intuitive reaction for the most part, rather than an objectively derived conclusion. Is it a causative relationship? Or is it no more than an associative connection? And what sort of conflict is the most frequent outcome—civil disorders within nations or hostile relations between nations? These considerations are the subject of this paper. (Because detailed author-itative publications are unusually scarce in sub-Saharan Africa, the pa-per concludes with a lenghty list ofpublications that the author has found specially useful.)

THE POPULATION FACTOR

The standard U.N. projections propose that sub-Saharan Africa's num-bers will surge from 508 million today to an ultimate total of more than 2 billion before zero growth is reached early in the 22nd century. This projection depends of course on large numbers of people finding the wherewithal to sustain themselves throughout a normal life-span. To that extent, strictly demographic projections are made in something of a natural-resource vacuum. Of course they are projections, *not* pre-dictions; still less are they forecasts. But they tend to be perceived as assertions of likely fact.

While this projection takes account of an expected decline in fertil-ity, which in turn is taken to reflect some degree of socioeconomic advancement, it is divorced, by its very nature, from considerations of the natural-resource base that sustains human communities of what-ever sort and condition. In particular it does not—indeed it cannot—take proper account of adverse trends in the natural-resource base that sustains many human activities in sub-Saharan Africa. It assumes that there will be some increase in human welfare (achieved in part through enhanced utilization of natural resources), leading to reduced fertility rates. But it does not recognize there can also be mis-use and over-use of natural resources, leading to little advance in human wel-fare, or even a fall-off in welfare—with, in turn, negative implications for fertility decline.

Even while sub-Saharan Africa has continued to produce infants at a rate far faster than anywhere else in the world, since 1960 the region has been growing poorer and hungrier in absolute as well as relative terms. Today's average per-capita income of $ 250 is only 95 percent of real income in 1960; of the 36 poorest countries in the world, 29 are in Africa (Chidzero, 1988; De Cuellar, 1988). Worse, the average per-capita agricultural production has declined by an average of 2 percent annu-ally since 1970 (Tables 3 and 4); and the World Bank (1988) estimates that production is unlikely to grow at more than 2.5 percent per year for at least the next two decades, even while population growth re-mains at 3 percent or more per year. Result, food output per head, which has declined by 20 percent since 1970, is scheduled to decline by a further 30 percent during the next 25 years (Dow, 1985; Hendry, 1988; United Nations World Food Council, 1988; U.S. Department of Agriculture, 1988; World Hunger Program, 1989). Thus the region serves as a prime example of an "adverse outlook" scenario (Sai, 1984; Goliber, 1985; Independent Commission on International Humanitarian Issues, 1985; World Bank, 1984 and 1986).

Of course population growth is not the only factor. Also to blame are the generally harsh environments of the region, its unreliable climate, its adverse trade terms, its weak infrastructure, and its faulty development policies (Timberlake, 1985). But there is much agreement that the list of problems is headed by population growth.

At the same time, there is a key question to be raised. Does the region still feature such uniquely high population growth rates? Not in Ethiopia, surely, where the death rate continues to be officially estimated at no more than 15 per 1000 per year, and where the growth rate is theoretically projected to climb from the present level of 3 percent to almost 4 percent by the year 2000. Nor, surely, do growth rates remain as high as formerly in at least a dozen other countries that have been suffering severe drought and famine. Of the population of almost 450 million people in 1985, 250 million were chronically malnourished, 150 million were subject to acute food deficits, and 30 million were actually starving (Independent Commission on International Humanitarian Issues, 1985). During 1986 and 1988 there has been some respite, thanks to better rains. But because of unpromising baseline conditions generally, and particularly in the way of harsh climate together with widespread ecological degradation and poor agricultural policies, the return of only moderately adverse weather conditions could quickly trigger a renewed onset of broadscale famine.

Those observers who consider this scenario is alarmist should remember that the region already imports one-fifth of its cereal requirements, at a cost that in current dollars rose seven-fold from $ 1.9 billion in 1970 to $ 12.4 billion in 1985 (Tables 3 and 4) (Mellor *et al.*, 1987; Yudelman, 1985; and for general background on the agricultural situation, see Chazan and Shaw, 1988; Gakou, 1987; Sanderson, 1984; Swaminathan and Sinha, 1986). In 1980 imports of food grains alone reached almost 21 million tons (plus food aid of another 1.5 million tons), enough to feed 135 million Africans or roughly one in three persons; these imports cost the region more than $ 5 billion in scarce foreign exchange. Yet food imports are projected to rise within 20 years to almost three times their 1980 level—even though the average percapita consumption of food grains is projected to remain at little more than 250 kg per year, by comparison with Asia and South America, around 300 kg each.

In terms of individual countries, as far back as 1975 there were 14 countries, accounting for one-third of the region's land area and one half of its total population, that featured populaces already too large in relation to their cultivable land to grow enough food for themselves, assuming continuation of subsistence farming methods (the assumption has been borne out by the stagnation of agricultural productivity in most of the region). Even if agricultural practices were to be upgraded with intermediate technology to reach commercial levels, half of the countries would still not manage to feed themselves from their own land by the year 2000 (Food and Agriculture Organization, 1984; Higgins *et al.*, 1984). Among these nations are Senegal, Mali, Sierra Leone, Togo, Benin, Burkina Faso, Malawi and Swaziland, none of which is expected to be generating enough goods for export to allow purchase of adequate food stocks abroad. A further lengthy list of countries will be unable to feed even half their projected year 2000 populations from their own land: Ethiopia, Nigeria, Uganda, Niger, Kenya, Lesotho, Burundi, Somalia and Rwanda—but fortunately Nigeria, Niger and Kenya could well be exporting enough goods to allow them to purchase food elsewhere (Food and Agriculture Organization, 1986).

True, a number of other countries in the region feature enough cultivable land to support theoretically larger populations than at present. But because their planning capabilities are not yet equal to the challenge of managing ultra-rapid growth, their populaces receive less and less to eat.

Humanitarian considerations apart, there is a security aspect to food shortages. Insufficient food eventually serves as a major source of frustration, civil disorder and outright violence. Already there have been food riots in urban communities of Ghana, Sudan, Ethiopia, Uganda, Angola, Zambia, Mozambique and Zimbabwe (Myers, 1986).

Fortunately there is much scope to expand food production, primarily through irrigation (Moris, 1987). Of irrigation potential, Sudan has developed only 52 percent, Zimbabwe 52 percent also, Kenya 14 percent and Tanzania 6 percent, while other countries reveal a similarly low level of development (Food and Agriculture Organization, 1985 and 1987). Not surprisingly, the largest need for irrigation water is during the dry season, when water availability can fall as low as 10 percent of annual flow. But in a region which is unduly dry, water deficits are already a pronounced problem, and likely to get worse fast. So let us take a quick look at this key issue.

WATER DEFICITS

As has been frequently pointed out (Falkenmark, 1986a and b; United Nations, 1987), sub-Saharan Africa is the driest region of the developing tropics. This fact has profound implications for the region's development outlook, especially for population pressures and agriculture (though remarkably enough, the point was scarcely recognized by the Brundtland Report). In particular, a large number of savannah countries are situated in areas that are marginal in terms of water supplies.

The problem lies not only with meagre amounts of rainfall in total (Solomon *et al*, 1987; World Meteorological Organization, 1987). It also relates to rainfall distribution around the year. Two-thirds of the region receives more than half its annual rainfall in just three months. This rainfall pattern therefore limits the length of the growing season, defined as that part of the year when there is enough water in the soil to satisfy at least half the water demands of growing crops. In turn this means that in as much as 44 percent of the region, drought is already a major limitation to rainfed agriculture (on top of the 18 percent where soil composition is a major limitation too). (Falkenmark, 1988 and 1989).

Many large human communities suffer from water deficits, whether in the form of water stress or outright scarcity of water. By way of comparison, they suffer these deficits on a scale equivalent at least to water problems now afflicting the lower Colorado Basin in the United States. By the year 2000 around 350 million people, or almost half the region's total, are projected to be living in water-stressed countries, and another 150 million in countries suffering from absolute scarcity of water (Falkenmark, 1987 and 1988). But well before then, and even were there to be high-level agronomic inputs in the form of advanced irrigation and water-efficiency systems, the waterstressed countries are projected to include Nigeria, Ethiopia, Somalia, Tanzania, Malawi, Zimbabwe and Lesotho, with a total of 212 million people today; by the year 2025 virtually all these countries are projected to suffer from absolute scarcity of water. Worse, within just a few years the water-scarce countries are projected to include Kenya, Rwanda and Burundi, and a good number of others are expected to be experiencing significant water stresses. Within a 5–35 year perspective, then, lack of sufficient water is likely to become a fundamental constraint on development for two-thirds of the region's population.

As we shall now see, all this has important implications for security concerns.

INTERNATIONAL RIVER BASINS

Tightening constraints on water supplies have profound implications for international river basins (Gleick, 1989). Of 200 sizeable river basins shared by two or more nations, 57 are in Africa. Among countries dominated by international river basins and with low per-capita water availability, a notable example is Ethiopia, with 80 percent of its area in the Nile river basin and per-capita water availability only 2.39×10^3 cu.ms. per year. Still worse placed is Sudan, with 81 percent of its area in the Nile river basin, and per-capita water availability a mere 1.31×10^3 (World Resources Institute, 1988).

This highlights the international signficance of the Nile's main watershed located in Ethiopia. The principal downstream nation, Egypt, faces growing water shortages. (True, Egypt is outside sub-Saharan Africa, but by virtue of the ecological linkages represented by the Nile, its welfare is closely connected to what goes on in the river's catchments.) Eight successive drought years in the watershed areas of Ethiopia (also in Uganda and other countries of equatorial Africa) reduced the Nile flows by mid-1988 to their lowest levels since 1913. During the first half of 1988 it seemed that storage water in Lake Nasser would prove enough only for the harvest of that year, whereafter the early-1989 planting season could be so curtailed through lack of irrigation water that Egypt would have to import a further 15 percent of its food needs through this cause alone (Stains, 1987). There would also have had to be a halt on Egypt's ambitious programmes to expand its irrigation network and to reclaim desertlands. Were the drier weather regimes in the upstream catchments to have persisted (as many climatologists anticipate in light of the green house effect; see e.g. Bradley *et al.*, 1987), the results would have been, according to Mr. Sarwat Fahamy, Director of the Nile Water Control Authority, "a calamity".

Moreover the water shortages would have affected not only agriculture. Low flows into the Aswan Dam's hydropower turbines, which supply 40 percent of Egypt's electricity needs, had already reduced power output by 20 percent. Another dry year in the catchment zones upstream would have resulted in a 60 percent power cut, at a cost of millions of dollars worth of lost energy each month. Since the deficit would have had to be made up with oil, the low-flows problem would have all but eliminated Egypt's oil-export revenues.

Hence the security significance of Ethiopia's plans to divert a greater snare of the Nile's waters (Starr and Stoll, 1988; Waterbury, 1987; Whittington and Haynes, 1985). Ethiopia controls the Blue Nile tributary, source of around 80 percent of Nile water entering Egypt; and Ethiopia has never joined Egypt and Sudan in a legal agreement to regulate the share-out of the Nile's waters. On the contrary, it has regularly asserted that as a sovereign state it feels at complete liberty to dispose of its natural resources in whatever manner best promotes the welfare of its people (Shahin, 1985). Ethiopia is resettling 1.5 million impoverished peasants from the degraded highlands into its fertile south western sector; and in order to supply irrigation water for the

217

new settlements, the government plans to divert up to 39 percent of the Blue Nile's waters before they leave the nation's territory. In 1985 Egypt's Foreign Minister Butros Ghali asserted "We depend upon the Nile 100 percent. The next war in our region will be over the waters of the Nile, not over politics....Washington does not take this seriously, because everything for the United States relates to Israel, oil and the Middle East."

THE POPULATION FACTOR REVISISTED

The considerations thus far have profound implications for population projections for the region, as well as for population-related planning and development strategies overall, and for prospects of stability and peace in this turmoil-ridden region. In particular, planners need to consider the environmental impacts of growing multitudes of impoverished peasants (Bilsborrow and Stupp, 1987; Hendry, 1988; World Bank, 1985). There is hardly any agent more destructive of key natural resources—notably soil cover, grasslands and forests—than the subsistence cultivator who, due to over-crowding, cannot gain his livelihood by cultivating in traditional farmlands. Often sidelined from the development process by virtue of inequitable political and socioeconomic factors, this marginal person is inclined to seek an alternative livelihood in marginal lands, i.e. lands that are too wet, too dry or too steep for conventional agriculture. Of course he is no more to be "blamed" for his actions than is a soldier to be held responsible for starting a war: the subsistence cultivator is often impelled by forces of political structures, economic systems and/or institutional factors, of which he may have little understanding and over which he exercises virtually no influence. But the result is the same: widespread deforestation, soil erosion and spread of deserts.

Another result is the growing incapacity of the natural-resource base to sustain ever-increasing numbers of people with a growing sense of desperation. Most significant of all for the present analysis, a still further result is apt to be the creation of first-rate breeding grounds for disaffection, disputes and revolts of various kinds.

ETHIOPIA AND THE HORN OF AFRICA

Ethiopia epitomizes those nations with high population growth rates that must work exceptionally hard already to satisfy their food needs. When a nation is impoverished to start off with, it may well find the task all but beyond its means. So the need grows faster still, until eventually it threatens the very structure and stability of the nation. In Ethiopia the essentials of everyday life, in terms of food supplies alone, are increasingly maintained by outside agencies rather than by the government. The country's output of cereal grain in 1980, 5.1 million tonnes, has been theoretically projected (World Bank, 1986) to expand to 7.3 million tonnes in 1990—though in fact, food production has been declining by an average of 1 percent per year during the 1980s. Yet despite this projected increase to 7.3 million tonnes of cereal-grain output in 1990, Ethiopia's need for cereal imports is projected jump from 214,000 tonnes in 1980 to more than 3 million tonnes by 1990.

As a consequence of the severe imbalance between population and food supplies, plus associated political upheavals, there are now at least 5.2 million people threatened today (late 1988), and a total of 14.7 million people, or nearly half the population, are chronically undernourished, i.e. they do not receive enough food to support an active working life. As a further result, there are now at least 3 million displaced people within Ethiopia, and another half million in refugee camps in Sudan (where they are explicitly recognized as environmental refugees rather than political refugees).

Let us briefly review the past record that has led to this situation. By the early 1970s, as much as 470,000 sq. kms. of Ethiopia's traditional farming areas in the highland zone were seriously or severely eroded, to the extent that they were losing an estimated 1 billion tonnes of topsoil per year (more recent, and more refined estimates (Hurni, 1985) put the loss at 3.5 billion tonnes per year). This massive soil erosion was due partly to rudimentary agricultural practices, partly to inequitable land-tenure systems, and partly to the pressures generated by a population that increased from 18 million in 1950 to 25 million in 1970. The results included a marked fall-off in agricultural production accompanied by food shortages in cities, with ensuing disorders that precipitated the overthrow of Emperior Hailie Selassie in 1974 (Farer, 1979; Shepherd, 1975; Tareke, 1977).

The new Dergue regime did not move fast enough to restore agriculture (Clarke, 1986; Clay and Holcomb, 1986; Eshetu and Teshome, 1984; Hancock, 1985; Jansson *et al.*, 1987). For this reason among others, throngs of impoverished peasants started to stream into the country's lowlands, including the Ogaden zone that borders Somalia—a zone of long-standing conflict between the two nations. In Somalia too, steadily increasing human numbers, together with inefficient agricultural practices, had led to much over-taxing of traditional farmlands. So for these reasons (plus some ethnic complications), there was a migration into the Ogaden from the Somali side as well. The result was a clash between the two sides, and an outbreak of hostilities in 1977 (Petrides, 1983).

The conflict was made worse by foreign interests (Selassie, 1980). The Horn of Africa lies adjacent to the oil-tanker lanes that extend from the Persian Gulf to the industrialized nations of the Western world. It also borders onto the Red Sea and the Indian Oceans, two areas that are becoming increasingly important in geopolitics. So the superpowers began to pour in support for their respective allies. Primarily as a result of this arms race, Ethiopia in 1981 spent $ 447 million on defense, and Somalia $ 105 million. Added to the outlays of previous years, the total sum expended in the Horn of Africa because of the Ogaden clash can be estimated at well over $ 1 billion during a five-year period (Luckham and Bekele, 1984). If only a small part of that sum had been allocated ahead of time to reforestation, soil safeguards and associated aspects of restoring the agricultural resource base in the two countries (estimated by the United Nations Anti-Desertification Plan to cost no more than $ 50 million a year), the disastrous outcome could well have been avoided.

Environmental breakdown and food shortages have now become endemic in Ethiopia (Gamachu, 1988). The mid-1980s drought has been no more than a triggering factor, precipitating a crisis that has been building up through the pressure of population growth and agricultural mismanagement.

While we can scarcely assert that population growth has been a prime or direct cause of the recent turmoil in Ethiopia, it has certainly helped to worsen the situation. The Dergue government, like the Selassie regime before it, has been unable to resolve the problem of growing human numbers with growing aspirations seeking to survive off a deteriorating natural-resource base. Even though Ethiopia has one of the lowest population growth rates in the Third World—no more than 2 percent per year for most of the time since 1950, whereas the average for much of sub-Saharan Africa is now 3 percent or more—this has still allowed Ethiopia's population to increase from 18 million in 1950 to 48 million today.

Consider what could have been achieved by population regulation. If Ethiopia had established a strong family-planning programme from 1970 onwards, and if the programme had achieved only moderate success, it might have prevented 1.7 million births by 1985. According to family-planning agencies, it would likely have cost around $ 170 million. Emergency food rushed to Ethiopia during the 1985 famine crisis amounted to 300,000 tonnes and supplied enough food to sustain 1.7 million people—at a cost of about $ 170 million (Ehrlich, 1985).

The region features other "flashpoints of conflict" in those several areas where fast-growing communities are pushing against the limits of their natural-resource base. Consider, for instance, Sudan, and some root causes of the 1985 coup that toppled the Nimeiry regime. Four main factors appear to have been at issue (Ahmed *et al.*, 1988; Ali, 1987; Cater, 1986). First was the civil war in the south, provoked by fears that water that "belongs" to the southerners was being diverted to the northerners. Second was a 500-percent rise in price of basic food staples as a result of drought and associated problems of declining agriculture. Third was a severe shortage of fuelwood in the north, after local supplies had been sorely depleted and the main supplementary supplies from the south had been cut off, precipitating a fast-growing demand for kerosene and petrol with price rises of more than 300 percent. Fourth was the mass migration into the Khartoum area for the Kordofan and Darfur Provinces, where gross deterioration of grazing lands had been exacerbated by drought. All of these factors had a strong environmental component—and all were exacerbated by the pressures of a population growing at 2.8 percent per year.

Other flashpoint countries include Chad and Mozambique, which, together with Ethiopia and Sudan, rank among the most under-developed nations of sub-Saharan Africa. All feature fast-growing populations; all suffer from severe environmental impoverishment; all have experienced prolonged political instability; all are subject to periodic military upheavals. Significantly, these are all nations with some of the highest per-capita expenditures on military activities.

ENVIRONMENTAL REFUGEES

In the wake of population pressures and environmental breakdown, a new phenomenon has emerged, that of "environmental refugees" (El Hinawi, 1985; Myers, 1986; United Nations Fund for Population Activities, 1988; and on link-ups with the broader refugee situation, see Clark, 1987). These are people who feel obliged to leave their home-lands because of declining means of livelihood, which in turn stems from over-rapid population growth and severe environmental degra-dation. The immediate cause may often appear in the form of military activity. But the underlying cause may have more to do with deterio-ration of the local natural-resource base and its incapacity to support the citizenry.

We have already noted the situation in Ethiopia. In Ivory Coast at least one-fifth of the population are unofficial immigrants from the Sahel, causing an acute problem of deforestation due to land hunger: onethird of landless people are immigrants. In Sudan there are now close to 1 million refugees from outside, adding to the economic burdens and political instability of that over-strained country—where almost another 1 million people have had to flee their homes for various environmental reasons, notably desertification. In both Ivory

Coast and Sudan the primary reason for the refugee phenomenon is desertification; and in the region as a whole, people displaced by drought were estimated to total at least 3 million in early 1986 (Glantz and Katz 1987; see also United Nations Environment Programme, 1985). By the year 2,000 there could well be an additional 70 million people affected by desertification (Dregne and Tucker, 1988).

Counting all refugees in the region, the total has grown from less than 1 million in 1970 to 3 million in 1980 and 10 million in 1985, or well over half of all refugees worldwide. This recent outburst is not surprising in a region where (as noted above) in 1985 some 150 million people, or one-third of the total, faced food shortages, and where 30 million suffered from famine: a result not so much of recent drought as of long-standing environmental degradation compounded by ultra-rapid population growth and faulty development policies. Although there are no accurate breakdown figures, this writer believes, on the basis of 24 years' residence in the region, that many more refugees deserve to be classified primarily as environmental refugees rather than as political refugees.

BETTER NEWS

Population

Fortunately there is some better news. On the population front, three-quarters of Africans now live in countries with governments that view their population growth rates as too high (Birdsall and Sai, 1988). As a measure of the recent progress in this respect, recall that only half of all Africans lived in such countries as recently as 1983. In accord with the Kilimanjaro Programme of Action on Population, which calls for the supply of family-planning services to all persons in question, many governments have begun to implement policies and programmes to tackle the problem. Despite the built-in momentum of past population growth, there is still much opportunity to reduce the long-term problem.

Were Nigeria, for example, to implement its new family-planning programme with appropriate vigour, it could conceivably achieve an average of two children per woman (replacement fertility) within 30 years. If this rapid fertility decline had begun in 1985, there would have been only 195 million Nigerians in the year 2015, and 270 million by the time that growth finally fades away a century hence. Since the rapid fertility decline cannot start before 1990 at best, the totals will expand to 219 million and 327 million; and if the fertility decline is delayed until 1995, then 239 million and 400 million. In other words, a delay of only

10 years from 1985 to 1995 will make a difference of 130 million people in the ultimate size of Nigeria's population. In 1982 when the country still had fewer than 100 million people, Nigeria felt so over-burdened that it summarily expelled between 2 and 3 million aliens, causing largescale human misery and generating acute tensions with neighbouring countries.

On a region-wide scale, if African women average two children each by 2030, the ultimate population will be 1.4 billion people. But if the two-child average is delayed until 2065, a gap of only 35 years, the ultimate figure will be more than three times greater, 4.4 billion.

To date only 3 or 4 percent of reproductive women practice contraception (compared with at least 30 percent in India and 70 percent in China). It costs around $ 20 to supply a contraceptive user for one year. So if 25 percent of at-risk women in the region are to be supplied by the year 2000, there will be an annual bill of $ 640 million. But thus far family-planning budgets amount to less than $ 100 million—and over half of that comes via foreign aid, constituting a bare 1 percent of all aid to the region. Even to restrict population growth to the median level projection of 1.45 billion people by 2025 will require an increase in contraceptive use to 59 percent of all at-risk women—or enough to bring down fertility rates from 6.4 to 3.2 children per woman (compared with all other developing countries, down from 4.0 to 2.3 children—this latter figure being the same as for the developed world in 1980). It will also mean that contraception-using women in the region must increase from 11 million in 1984 to 168 million, an increase of 1,404 percent (contrast Southeast Asia, only 279 percent, Latin America 184 percent and East Asia 30 percent).

Fortunately, there is a success story to point the way ahead. From 1982 the Zimbabwe government has been using an array of tax incentives and other inducements for birth-control measures. As a result, Zimbabwean women have increased their rate of contraceptive use from 14 percent in 1982 to 38 percent in 1987, one of the most remarkable achievements of this sort in the world.

Environment and agriculture

On the environmental front too, many new initiatives are underway to reduce soil erosion, desertification and deforestation. In conjunction with policy reforms in the agricultural sector such as price incentives, reduced subsidies, better extension services and expanded marketing networks, the result has been an increase in agricultural production by 3 percent in 1986 (World Bank, 1988). Zimbabwe has shown what is possible by turning from a net food importer as recently as 1982 into a net food exporter in 1987.

Sub-Saharan Africa and the North

But future progress along these lines depends in major measure on the international economic climate. The outlook is not promising. In 1986 commodity prices hit a 30-year low point, reducing export revenues by 30 percent below the previous year's level; in sub-Saharan Africa declining raw-materials prices eliminated $ 19 billion from the region's revenues, or about four times the amount which the region was promised in emergency aid during that drought-stricken year (Grant, 1989).

Still worse is the debt problem. Between 1982 and 1987 the region's debt increased by 78 percent (compare Asia 65 percent and Latin America 30 percent). Debt-service payments rose from 15 percent of export earnings in 1980 to 31 percent in 1986. The region now owes more than $ 200 billion to the West. The 17 most indebted countries face annual payments of $ 6.9 billion, three times more than in 1985. Many African countries now spend more than 50 percent of their export earnings on debt interest payments; some must divert virtually all their export earnings to this end. Not surprisingly, in 1987 net payments were zero. These economic straits make it increasingly difficult for the region to attract development capital from the advanced world, further frustrating its efforts to redress poverty and to meet basic needs. Together with decreased financial flows from international lenders, the upshot has been a net decline in resource transfers from an estimated $ 10 billion in 1982 to $ 1 billion in 1985, and to $ 0.95 billion in 1987.

To make the debt problem worse, affluent nations have largely failed to deliver on their promises of increased aid and more equitable terms of trade. Development assistance to the region actually declined slightly in 1986, and private lending has continued to plunge. Official development assistance from OECD nations rose hardly at all from 1980 to 1987 (Lancaster, 1988), while longterm non-concessional loans were halved (Grant, 1989). In 1985, for every dollar the West donated to famine-stricken African nations, it took back two dollars in debt repayments; whereas between 1985 and 1986 total Third World debt increased by almost 10 percent (by approximately $ 70 billion), in sub-Saharan Africa it soared by almost 20 percent (roughly $ 25 billion). By 1995 the region's debt-service obligations could well reach $ 45 billion—altogether impossible sum.

Fortunately there is a bright star in this dark picture. Canada has recently forgiven outstanding public debt of the 13 most impoverished nations in the region, worth a total of $ 586 million. Of all the region's debt, some 80 percent is publically held, meaning there is great scope for similar creative initiatives on the part of other creditor nations. Were other donor nations to follow Canada's lead and convert their outstanding development assistance loans into grants, it would save the region between $ 200 and $ 300 million a year.

CONCLUSION

These, then, are some major factors in the complex of problems that interlink population growth, environmental degradation, economic decline and growing political instability, culminating in a spread of conflict and violence. The outlook is far from positive: recall the two quotations at the head of this paper, by two leading figures in the region. Without greatly increased ameliorative measures both within the region and outside, the prospect must be a continuing deterioration in the security situation, with repercussions that will hardly be confined to sub-Saharan Africa.

REFERENCES

Adedegi, A. 1983: *The Economic Commission for Africa and Africa's Development 1983–2008*. Economic Commission for Africa, Addis Ababa, Ethiopia.

Ahmed, A.R.A.Z. and 18 others. 1988: *War Wounds: Development Costs of Conflict in Southern Sudan*. The Panos Institute, London, U.K.

Ali, A.A. 1987: *Sudan Economy in Disarray*. Ithaca Press, London, U.K.

Bilsborrow, R.E. and P. Stupp. 1987: *Demographic Effects on Rural Development in Sub-Sahara Africa: An Assessment of the Literature and Recommendations*. International Labour Organization, Geneva, Switzerland.

Birdsall, N. and F.T. Sai. 1988: Family Planning Services in Sub-Saharan Africa. *Finance and Development* (March): 28–31.

Bradley, R.S., H.F. Diaz, J.K. Eischeid, P.D. Jones, P.M. Kelly and C.M. Goodess 1987: Precipitation Fluctuations Over Northern Hemisphere Land Areas Since the Mid-19th Century. *Science* 237: 171–175.

Brown, L.R. and E.C. Wolf. 1985: *Reversing Africa's Decline*. Worldwatch Institute, Wasington D.C.

Cater, N. 1986: *Sudan: The Roots of Famine*. Oxfam Publications, Oxford, U.K.

Chazan, N. and T.M. Shaw, eds. 1988: *Coping With Africa's Food Crisis*. Lynn Rienner Publishers, Boulder, Colorado, U.S.A.

Chidzero, B.T. 1988: Africa and the World Economy. *World Futures* 25: 157–162.

Clark, L. 1987: *Country Reports on Five Key Asylum Countries in Eastern and Southern Africa*. Refugee Policy Group, Washington D.C., U.S.A.

Clarke, J. 1986: *Resettlement and Rehabilitation: Ethiopia's Campaign Against Famine.* Gollancz Publishers, London, U.K.

Clay, J.W. and B.K. Holcomb. 1986: *Politics and the Ethiopian Famine 1984–85.* Prentice-Hall Publishers, Engelwood Cliffs, New Jersey, U.S.A.

De Cuellar, P. 1988: *Africa's Economic Situation.* Office of the Secretary General, United Nations, New York, U.S.A.

Dow, M.M. 1985: Food and Security. *Bulletin of the Atomic Scientists* 41: 21–26.

Ehrlich, A.H. 1985: Critical Masses. *The Humanist* 45: 18–22, 36.

El Hinawi, E. 1985: *Environmental Refugees.* United Nations Environment Programme, Nairobi, Kenya.

Eshetu, C. and M. Tesnome. 1984: *Land Settlement in Ethiopia: A Review of Developments.* Addis Ababa University Press, Addis Ababa, Ethiopia.

Falkenmark, M. 1986a: Macro-Scale Water Supply/Demand Comparison on the Global Scene. *Beitrage zur Hydrologie* 6: 15–40.

Falkenmark, M. 1986b. Fresh Water—Time for a Modified Approach. *Ambio* 15: 192–200.

Falkenmark, M. 1987: Water-Related Constraints to African Development in the Next Few Decades. In *Water for the Future: Hydrology in Perspective* (IAHS publication) 164: 439–453.

Falkenmark, M. 1988: *The Massive Water Penury Now Threatening Africa—Why Isn't It Addressed?* Stockholm Group for Studies on Natural Resources Management, Stockholm, Sweden.

Falkenmark, M. 1989: Water, Not Land—Long-Term Obstacle to Food Production in Africa. In D. Pimentel and C.W. Hall, eds., *Food and Natural Resources,* Academic Press, New York, U.S.A. (in press).

Farer, T.J. 1979: *War Clouds on the Horn of Africa.* Carnegie Endowment for International Peace, Washington D.C., U.S.A.

Farmer, G. and T.M.L. Wigley. 1985: *Climatic Trends for Tropical Africa.* Climatic Research Unit, University of East Anglia, Norwich, U.K.

Food and Agriculture Organization. 1984: *Land, Food and People.* Food and Agriculture Organization, Rome, Italy.

Food and Agriculture Organization. 1985: *Estimates of Areas Developed: Irrigation Potentials.* Land and Water Division, Food and Agriculture Organization, Rome, Italy.

Food and Agriculture Organization. 1986: *African Agriculture, the Next 25 Years: The Land Resource Base.* Food and Agriculture Organization, Rome, Italy.

Food and Agriculture Organization. 1987: *Irrigation and Water Resources Potential of Africa.* Food and Agriculture Organization, Rome, Italy.

Food and Agriculture Organization. 1988: *State of Food and Agriculture 1987.* Food and Agriculture Organization, Rome, Italy.

Gakou, M.L. 1987: *Crisis in African Agriculture.* Zed Books, London, U.K.

Gamachu, D. 1988: *Environment and Development in Ethiopia.* International Institute for Relief and Development, Geneva, Switzerland.

Glantz, M.H., (ed.) 1987: *Drought and Hunger in Africa.* Cambridge University Press, Cambridge, U.K.

Glantz, M.H. and R.W. Katz. 1987: African Drought and Its Impacts: Revived Interest in a Current Phenomenon. *Desertification Control Bulletin* 14: 22–23

Gleick, P.H. 1989: The Implications of Global Climatic Changes for International Security. *Climatic Change* (in press).

Goliber, T.J. 1985: *Sub-Saharan Africa: Population Pressures on Development.* Population Reference Bureau, Washington D.C., U.S.A.

Grant, J.P. 1989: *The State of the World's Children 1989.* UNICEF, New York, USA.

Gulhati, R. 1988: *The Political Economy of Reform in Sub-Saharan Africa*. Economic Development Institute, The World Bank, Washington DC, USA.

Hancock, G. 1985: *Ethiopia: The Challenge of Hunger*. Gollancz Publishers, London, U.K.

Harrison, P. 1987: *The Greening of Africa*. Palladin Books, London, U.K.

Hendry, P. 1988: *Food and Population: Beyond Five Billion*. Population Bulletin 43 (Population Reference Bureau, Washington D.C., U.S.A.).

Higgins, G.M., A. Kassam, L. Naiken, G. Fischer and M. Shah. 1984: *Potential Population Supporting Capacities of Lands in the Developing World*. Food and Agriculture Organization, Rome, Italy.

Hurni, H. 1985: *Erosion, Productivity, Conservation Systems in Ethiopia*. Soil Conservation Research Project, University of Berne, Berne, Switzerland.

Independent Commission on International Humanitarian Issues, 1985: *Famine: A Man-Made Disaster*. Pan Books, London, U.K.

Jansson, K., M. Harris and A. Penrose. 1987: *The Ethiopian Famine*. Humanities Press International, Atlantic Highlands, New Jersey, U.S.A.

Lancaster, C. 1988: *US Aid to Sub-Saharan Africa: Challenges, Constraints and Choice*. Center for Strategic and Interantional Studies, Washington DC, USA.

Lassoie, J.P. and S. Kyle. 1989: *Policy Reform and Natural Resources Management in Sub-Saharan Africa*. Departments of Natural Resources and Agricultural Economics, Cornell University, Ithaca, NY, USA.

Lewes, L.A. and L. Berry. 1988: *African Environments and Resources*. Allen and Unwin, London, U.K.

Luckham, R. and D. Bekele. 1984: Foreign Powers and Militarism in the Horn of Africa. *Review of African Political Economy* 30: 820 and 31: 7–28.

Marcum, J.A. 1989: Africa: A Continent Adrift. *Foreign Affairs* 68: 159–179.

McNamara, R.S. 1985: *The Challenges for Sub-Saharan Africa*. Sir John Crawford Memorial Lecture, Washington D.C., U.S.A.

Mellor, J.W., C.L. Delgado and M.J. Blackie, eds. 1987: *Accelerating Food Production in Sub-Saharan Africa*. Johns Hopkins University Press, Baltimore, Maryland, U.S.A.

Moris, J. 1987: Irrigation as a Priority Solution in African Development. *Development Policy Review* 5: 99–123.

Myers, N. 1986: The Environmental Dimension to Security Issues. *The Environmentalist* 6(4): 251–257.

Myers, N. 1987: Population, Environment and Conflict. *Environmental Conservation* 14(1): 15–22.

Obasanjo, 0. 1987: *Issues in Agricultural Development in Africa*. The Hunger Project, New York, N.Y., U.S.A.

Petrides, S.P. 1983: *The Boundary Question Between Ethiopia and Somalia*. New Social Press, New Delhi, India.Population Reference Bureau. 1988. *World Population Data 1988*. Population Reference Bureau, Washington D.C., U.S.A.

Revelle, R. 1988: *Food, Population and Conflict in Africa*. Institute of Policy Studies, University of California, San Diego, California, U.S.A.

Population Reference Bureau. 1988: *World Population Data 1988*. Population Reference Bureau, Washington DC, USA.

Sai, F.T. 1984: *Population Factor in Africa's Development Dilemma*. Science 226: 801–805.

Selassie, W.B. 1980: *Conflict and Intervention in the Horn of Africa*. Monthly Review Press, New York, U.S.A.

Shahin, M. 1985: *Hydrology of the Nile Basin*. Elsevier Publishers, New York, U.S.A.

Shepherd, J. 1975: *The Politics of Starvation*. CarnegieEndowment for International Peace, Washington D.C., U.S.A.

Solomon, S.I., M. Beren and W. Hogg, eds. 1987: *The Influence of Climate Changes and Variability on the Hydrological Regime and Water Resources*. International Association for Hydrological Studies Publ. No. 168.

Stains, E.D. 1987: *Irrigation Briefing Paper*. Office of Irrigation and Land Development, U.S. AID, Cairo, Egypt.

Starr, J.R. and D.C. Stoll, eds. 1988: *U.S. Foreign Policy on Water Resources in the Middle East*. Westview Press, Boulder, Colorado, U.S.A.

Tareke, G. 1977: *Rural Protest in Ethiopia 1961–1970: A Study of Rebellion*. Unpublished doctoral dissertation, Syracuse University, Syracuse, New York, U.S.A.

Timberlake, L. 1985: *Africa in Crisis: The Causes, the Cures of Environmental Bankruptcy*. Earthscan Publications Ltd., London, U.K.

United Nations. 1987: *Water Resources: Progress in the Implementation of the Mar de Plata* Action Plan. Economic and Social Council, United Nations, New York, U.S.A.

United Nations Environment Programme. 1988: *Desertification Control in Africa: Actions and Directions of Inistitutions*. United Nations Environment Programme, Nairobi, Kenya.

United Nations Fund for Population Activities. 1988: *State of World Population Report*. United Nations Fund for Population Activities, New York, U.S.A.

United Nations High Commission for Refugees. 1987: *Refugees and Displaced Persons from Mozambique*. United Nations, New York, U.S.A.

United Nations World Food Council. 1988: *The Global State of Hunger and Malnutrition: 1988 Report*. United Nations World Food Council, New York, U.S.A.

U.S. Department of Agriculture. 1988: *World Grain Situation and Outlook*. Foreign Agricultural Service, U.S. Department of Agriculture, Washington D.C., U.S.A.

Vohra, B.B. 1987: *When Minor Becomes Major: Problems of Ground Water Management*. Advisory Board on Energy, New Delhi, India.

Waterbury, J. 1987: Legal and Institutional Arrangements for Managing Water Resources in the Nile Basin. *Water Resources Development* 3: 92–103.

Whittington, D. and K.E. Haynes. 1985: Nile Water for Whom? Emerging Conflicts in Water Allocation for Agricultural Expansion in Egypt and Sudan. I. P. Beaumont and K. McLaughlin, (eds.) *Agricultural Development in the Middle East*. John Wiley and Sons, New York, U.S.A.

World Bank 1983: *Sub-Saharan Africa: A Progress Report on Development Prospects and Programs*. World Bank, Washington D.C., U.S.A.

World Bank 1984: *Toward Sustained Development in Sub-Saharan Africa*: A Joint Program of Action. World Bank, Washington D.C., U.S.A.

World Bank 1985: *Population Growth and Policies in Sub-Saharan Africa*. World Bank, Washington D.C., U.S.A.

World Bank 1986: Financing Adjustment with Growth in sub-Saharan Africa, 1986–90. World Bank, Washington D.C., U.S.A.

World Bank 1987: *World Development Report 1987*. World Bank, Washington D.C., U.S.A.

World Bank 1988a: *The Challenge of Hunger in Africa: A Call to Action*. The World Bank, Washington DC.

World Bank 1988b: *World Development Report 1988*. World Bank, Washington D.C.

World Hunger Program. 1989: *Beyond Hunger: An African Vision of the 21st Century*. World Hunger Program, Brown University, Rhode Island, U.S.A.

World Meteorological Organization. 1987: *Water Resources and Climatic Change: Sensitivity of Water-Resource Systems to Climate Change and Variability.* World Climate Programme, World Meteorological Organization, Geneva, Switzerland.

World Resources Institute. 1988: *World Resources Report 1988.* World Resources Institute, Washington D.C., U.S.A.

Yudelman, M. 1985: Sub-Saharan Agricultural Research. *Bulletin of the Atomic Scientists*: 41(8): 35–38.

TABLES

Table I. *Sub-Saharan Africa population,   actual and projected (population in millions)*

Selected Countries	1950	1980	2000	2025	2050	2100	Total fertility rate 1988	Year of replacement level fertility	Growth rate 1988
Western Africa									
Ghana	4.4	11.5	23	40	53	62	5.8	2025	3.1
Ivory Coast	2.8	8.4	17	29	38	45	6.7	2030	3.1
Mali	3.3	6.7	11	20	28	35	6.7	2035	2.9
Niger	2.9	5.5	11	20	29	38	7.1	2040	2.9
Nigeria	40.6	84.7	163	295	412	509	6.6	2035	2.9
Eastern Africa									
Burundi	2.5	4.1	7	13	18	23	6.6	2035	2.9
Ethiopia	18.0	37.7	64	106	142	173	7.0	2035	3.0
Kenya	5.8	16.6	37	69	97	116	8.0	2030	4.1
Mozambique	6.5	12.1	22	39	54	67	6.1	2035	2.6
Rwanda	2.1	5.1	10	20	30	39	8.5	2040	3.7
Somalia	2.1	4.7	9	16	23	30	7.4	2040	3.1
Sudan	9.2	18.9	33	58	79	98	6.5	2035	2.8
Tanzania	7.9	18.8	37	69	96	120	7.1	2035	3.6
Uganda	4.8	12.6	25	46	64	80	7.0	2035	3.4
Zambia	2.4	5.6	11	20	2?	32	7.0	2030	3.7
Zimbabwe	2.4	7.0	14	25	33	38	6.5	2025	3.5
Middle Africa									
Angola	4.1	7.6	13	23	32	41	6.4	2040	2.6
Cameroon	4.6	8.7	17	30	42	50	5.9	2030	2.6
Zaire	14.2	27.1	50	86	116	139	6.1	2030	3.0
	140.6	303.4	574	1024	1413	1735			
Other Sub-Saharan Africa	31.8	59.6	104	178	245	306			
Total Sub-Saharan Africa	172.4	363.0	678	1202	1658	2041			

Source: McNamara, 1985; Population Reference Bureau, 1988; World Bank, 1985 and 1988.

Table II. *Population growth rates, by region, 1950-85 (annual average in percent)*

Region	1950-55	1960-65	1970-75	1975-80	1980-85
Africa	2.11	2.44	2.74	3.00	3.01
Latin America	2.72	2.80	2.51	2.37	2.30
East Asia*	2.08	1.81	2.36	1.47	1.20
South Asia	2.00	2.51	2.44	2.30	2.20
All developing countries	2.11	2.30	2.46	2.14	2.02

* excluding Japan

Source: McNamara, 1985.

Table III. *Sub-Saharan Africa: growth rates of per capita food production and levels of food imports*

Selected countries	Growth of per capita food production (%) (millions of USD)			Levels of food imports			
	1961-70	1971-79	1980-84	1961	1970	1980	1985
Western Africa							
Ghana	0.8	-3.5	-0.8	61.4	77.2	131.6	214.7
Ivory Coast	4.8	0.4	-1.5	37.8	82.2	486.5	422.2
Mali	-0.3	-0.8	-1.6	8.3	17.2	70.0	81.8
Niger	-1.7	0.0	-5.7	3.0	8.9	82.0	44.6
Nigeria	0.2	-0.7	-1.6	90.1	127.2	2085.4	1523.9
Eastern Africa							
Burundi	-0.3	0.0	-3.0	1.2	3.9	29.6	23.6
Ethiopia	2.4	-1.6	-3.9	7.9	16.8	105.9	158.4
Kenya	0.1	-1.6	-2.0	40.6	49.7	213.5	152.9
Mozambique	0.9	-3.6	-3.9	21.2	37.3	114.0	84.0
Rwanda	2.6	0.4	-1.2	0.3	4.1	43.7	39.7
Somalia	0.5	-4.6	-4.1	12.4	16.4	147.3	163.6
Sudan	2.2	0.4	-3.6	41.4	65.3	390.0	202.6
Tanzania	2.6	-0.7	-3.4	31.0	32.4	164.8	100.7
Uganda	0.7	-1.2	0.7	14.6	21.3	45.0	17.6
Zambi	-0.2	-1.9	-2.4	19.7	47.9	144.8	67.6
Zimbabwe	2.3	-2.8	-7.9	12.2	10.8	61.9	40.4
Middle Africa							
Angola	1.0	-2.2	-2.2	25.4	55.7	267.6	211.1
Cameroon	1.4	-0.5	-2.1	13.6	31.1	130.9	151.9
Zaire	1.3	-1.5	0.6	27.4	63.4	165.5	146.8

Source: World Bank, 1986 and 1987.

Table IV. *Growth rates of per capita food production and levels of food imports*

Country Groups	Growth of Per Capita Food Production (%)			Levels of Food Imports (millions of US$)		
	1961-70	1971-80	1980-84	1970	1980	1985
Sub-Sahara: low income	1.3	-1.3	-2.1	562	2,307	1,966
Sub-Sahara: other	0.6	-1.1	-1.8	554	4,200	3,353
Sun-Sahara: total	1.0	-1.2	-2.0	1116	6,507	5,319
Other Africa	0.8	-1.5	-1.8	770	7,382	7,129
Total Africa	0.9	-1.3	-1.9	1886	13,889	12,448

Source: World Bank, 1986 and 1987.

Environmental Stress and Security in Southern Africa

Adolfo Mascarenhas

INTRODUCTION

The Southern African region represents one of the few examples in the world where a political crisis has definitely led to gross stress to the environment and to human populations over large areas. Politically, the system of apartheid makes the area one of the most volatile parts of the world. By practicing apartheid, a white minority regime in power in the Republic of South Africa has denied basic political, economic and human rights to the African majority. Apartheid not only affects the inhabitants directly under the regime in South Africa itself, but with collaborators, violence has also spilled in neighbouring countries. The security of more than three million individuals has already ceased to exist, and the lives of hundreds of thousands of others are threatened. Large areas in Mozambique and Angola have been made uninhabitable and the two countries fight for their survival.

To meet the threat of apartheid, the nine majority-ruled countries in the region formed the Southern Africa Development Coordinating Conference (SADCC) in 1980. It has evolved into a unique instrument for the development of nations in the region and for the promotion of peace. Intensified cooperation and increased solidarity within SADCC not only strengthen individual states but have also led to innovative ways of tackling problems. Annual consultations and conferences have not solved problems; they have helped to identify problems so that they could be solved elsewhere (Mmusi, 1985).

ROOT CAUSES AND THE CONTEXT OF ENVIRONMENTAL STRESS

Notwithstanding the critical problems brought about by apartheid, it must also be recognized that a colonial legacy, poverty and severe economic difficulties brought about by internal and external factors have all contributed to the problem of ecological stress and insecurity in Southern Africa. Countries have been affected differently by these factors.

233

It appears that periods of major economic and political upheavals also bring in their wake environmental and resource stress. Some of the major attention given to the need for conservation in Africa took place before and during the period of the Great Depression prior to World War II. The present very severe economic difficulties prevalent in many African countries manifests itself on the environment. To survive, people resort to extracting the maximum from nature. For instance, the shock of the increasing oil prices after 1973 meant that right across Africa many rural and urban people simply could not afford kerosene and therefore went back to wood-based energy. But as has been demonstrated recently concerning bioenergy issues in Africa, before we can have effective policies and projects, it is essential that we first understand what the problems really are. (Leach G, Mearns R, 1988).

Secondly, although many conservationists continue to believe that they are dealing with apolitical issues, the reality is different. To the majority of Africans, most of whom live in rural areas, conservation is a political issue. Any intervention which alters the relationship between people and the environment critically effects their subsistence economy—critical because it affects both the dwelling place and the basic means of production. Fortunately, there is a growing body of literature which has begun to refute conventional notions about a political conservation in Africa (Anderson and Grove, 1988). The World Commission on Environment and Development, as will be elaborated later, similarly focuses on the political aspects of sustainability rather than its technical issues. (Brundtland, 1987). In this regard, it is at present the most relevant document for issues pertaining to conservation and development.

Thirdly, a rapidly growing population very heavily dependent on natural resources makes it fundamental that ecological considerations should be in the fore-front of development. However, the fact that ecological factors have not been given prominence can be attributed to past development policies and assumptions made about natural resources utilization. While it is tempting on this issue to directly relate population numbers to environmental stress, the real relationship is much more complex. Even when attempts are made to link population numbers with food availability, there is growing evidence that the root cause of famines lies elsewhere (Sen A, 1981). Indeed, increased food security at the household level can be achieved by focusing attention on women.

A fourth aspect is simply that environmental changes can take place irrespective of human intervention. For example, earthquakes, some of the consequences of floods, and droughts all belong to a different class of ecological stress. In this respect, if we are experiencing climate change, it is high time that African governments paid more attention to the implication of these changes. Ecological stress can take many forms and can have its origin outside human influences. For instance, the reduction of vegetation cover in an area can occur because of drought, as has been the case in many parts of Botswana. However, in most countries, almost imperceptibly the major transformation and even the disappearance of vegetation can be attributed to human action.

Generally, socio-political factors are both the cause and effect of ecological stress. National insecurity, like environmental stress, takes many forms. It can occur in segments of the population while there is stability at the national level. A central measure of insecurity is that pertaining to food. Although food security has been the focus of a great deal of attention in Southern Africa, including SADCC, it will not directly be a focus of this paper.

Below the national level, insecurity usually occurs among communities which for one reason or another have been marginalised. The case of the Ngorongoro Conservation Area depicts the insecurity of conservation without a human face. While attempts have been made to resolve the conflicts, there is no evidence of a solution. A persistent background has been the political will expressed by the state to tackle environmental issues. The recent attempts by Mozambique and Botswana to incorporate the environmental and resource dimension in their development will be the main body of this paper. However, to give a proper perspective to changes, it is essential that the environmental parameters be treated first.

THE PHYSICAL ASPECTS OF ECOLOGICAL STRESS

The countries of the Southern Africa region are extremely varied in size and physical attributes. Individual countries may face problems brought about by the environment; collectively they stand a far better chance of minimising environmental stress and ensuring the security and well being of their people. For instance, both Swaziland and Lesotho together cover less than 50,000 sq. kms., in contrast with the areal giant, Angola, which has an area of over 1.2 million sq. kms. Collectively, the nine countries comprising SADCC cover an area of more than 4.9 million sq. kms. and have a present population in excess of 70 million people.

Ecologically, too, the area is one of major contrasts. Thus, the snow-covered peak of Kilimanjaro in the tropics is the highest mountain in Africa, and Lake Tanganyika represents one of the deepest floors below sea level. The average altitude of Lesotho, at over 1200 m a.s.l., makes it the highest country in the world and therefore, its vegetation and aquatic life need to be better understood.

At this stage it is also important to underscore that overall climatic variation is one of the very major stresses experienced by nearly all countries.

In this brief paper it will not be possible to entirely give a coverage of environmental stress in all the SADCC countries. There is little systematic information or data available on the extent of current natural resources deterioration. Notwithstanding this fact, certain areas of stress can be identified as critical.

The arid and semi-arid areas

The single most important concern is that in nearly all SADCC countries there is much more degradation of land and other resources in the arid and semi arid areas than in other areas.

None of the countries has the same degree and extent of aridity experienced by Botswana. Only in the northeast and along the southeastern border is the average precipitation in excess of 500 mm p.a. Rainfall varies both annually, areally and seasonally. The lower the rainfall, the more erratic it becomes. Monthly coefficients of variation range from 60–114% for the Gaborone station (Bharlotra, 1985). Drought is ever present and the country is now recovering from a six-year drought. Temperatures may exceed 40°C, so that potential evapo-transpiration can be over 2000 mm per annum.

Lesotho, having some of the highest mountain ranges in Africa, frequently suffers from sudden snow storms and blizzards and also experiences drought in the lower areas of the south.

Nearly a third of Tanzania is climatically classified as semi–arid with an average rainfall of 500–700 mm. Severe eroding has taken place in several areas of Dodoma, Singida and parts of Iringa. The classical work on soil erosion and sedimentation undertaken in the late 1960s has increased quantitative data (Berry/Rapp/Temple). In five catchment areas which were studied in Dodoma, the rates of erosion were as much as 226–783 tons/km and denudation rates of as much as 10 mm/year have been recorded (Christiansson C, 1981). Several detailed studies have been undertaken since then.

Even more people are affected by drought conditions in Tanzania than in Botswana. There are several factors which account for this situation in Tanzania. The chief among these is the expansion of people into marginal lands, increasing cattle numbers and inadequate adaption of agricultural practices to the newly settled areas. Inadequate planning and preparation during villagisation period has also caused severe localized stress. Nevertheless, both Tanzania and Botswana in their different ways reflect how drought-induced food insecurity can be minimised.

Ecological areas experiencing stress

There are three ecological areas also experiencing stress: the mountains, coasts and wetlands. The mountain areas have been favourable areas for settlement since time immemorial. Highlands closer to the equator represent a much healthier and richer environment beyond a certain altitude, but not in the extreme south of our region, such as in Lesotho, where mountain situations are a problem. In Tanzania population densities are dense in most of the highland areas. Where natural vegetation has been systematically destroyed, as on parts of the Usambaras and Pare mountains, run off and soil erosion have increased greatly. In contrast, part of Kilimanjaro continues to hold increasing population through the incorporation of conservation-oriented agriculture.

The physical environments of the SADCC countries are heterogeneous. While degraded land-scapes may look similar, the processes involved are different. At one stage, it was envisaged that problems of land degradation could be easily resolved through technical solutions. It is now recognised that the problems are much more complex and cannot be resolved without the sound understanding and inclusion of the human dimension. Experiences of participation demonstrate that there are several options from which to choose.

In recent years there has been sufficient concern about the threat of desertification and erosion for several countries to take their own initiatives and arrest environmental deterioration. In both Zambia and Zimbabwe, National Conservation Strategies have been launched. At the SADCC level a special coordination unit for soil and water conservation and a land utilization programme has been established at Maseru, Lesotho. Among its many relevant projects is Erosion Hazard Mapping (Stocking M).

The coastal areas of both Tanzania and Mozambique show evidence of stress brought about by changes to the habitat. Stress is indicated by rapid erosion and encroachment of the sea, leaving bare exposed rock, sedimentation of bays, etc. Destruction of coral by human activity is one reason for erosion, but the actual causes of rapid erosion in recent years have yet to be fully understood. In addition, the destruction of mangrove through pollution or clearance leads to loss of much sea life (Mainoya and Siegel 1987).

Conservation is fundamental to sustainable development. As defined in the World Conservation Strategy, conservation is "managing the use of environment and natural resources to ensure the maximum sustainable benefits for present and succeeding generations".

THE SCRAMBLE FOR RESOURCES—INSECURITY AT A COMMUNITY LEVEL

The scramble for resources in Africa takes many forms. This scramble is best epitomised in the struggle between land use by pastoralists and those of "conservationists". The outcome of the struggle is that large areas of rangeland have been expropriated either in the form of national parks or for exclusive wildlfe use (Hjort, 1982). In Tanzania the Ngorongoro Conservation Area (NCA) presents a classical case of the conflicting interests between these two camps.

The NCA, centred around the Ngorongoro crater, covers an area of about 8,400 sq. kms. Archaelogically, the renowned Olduvai Gorge has been the home of some of the earliest-known human beings, evidenced by the remains of *Zinjanthropos* and *Homo habilis* dating back to over 1.75 million years (Leaky, 1981). There is a profusion of evidence in the form of grave cairns and stone works that a Stone Bowl people inhabited the area dating back at least 2,500 years. The present residents, the Maasai, came into the area about 150 years ago. Over a period of time they covered an area from the shores of Lake Victoria to the Indian Ocean in the East. However, during German and British rule, the area under their suzerainty became very reduced.

The present pastoral residents of the NCA number over 22,000 people. Many of the inhabitants consist of Maasai who were ordered to leave the Serengeti and Tarangire when these two areas were declared as national parks. The highlands of Ngorongoro with its lush mountain pasture were seen by the pastoralist as the key to the continued occupation of the surrounding dry plains.

The Maasai have attempted to be a self-providing society in an environment of great risks. Their food supply system is dependent on livestock. Because of marked changes in the pattern and amount of rainfall and high local variability, they have to move frequently in order to obtain water and pasture for their animals. To the Maasai this is the only least-cost option available. The Maasai grazing grounds are frequently also areas of large and varied wildlife populations.

However, since generations of Maasai have coexisted with wildlife, they have the knowledge, skills and courage to continue a tradition of conserving animals—wild as well as domesticated. Unfortunately, there are conflicting interests. The pastoralists in the NCA are under intense pressure from some quarters in the conservation circles, state enterprises and those seeking "empty" lands. But the greatest threat is simply that there is a drastic reduction in the number of livestock for the Maasai to sustain their traditional life style (see Table I).

Managing conflicts

The initial problem arose with the single land use concept of national parks. This first made it incompatible to have the Maasai and their livestock as well as wildlife in the proposed Serengeti National Park (SNP). An attempt to reduce the size of the SNP, so that pastoralists could have access to land, caused a great deal of concern among conservationists. including powerful individuals in Europe and North America. As a result, in 1959 the eastern part of the SNP was excised and the Ngorongoro Conservation Area Authority (NCAA) was created.

The NCAA was charged with ensuring that there would be multiple use of land so as to assist in "Conserving and Developing the Natural Resources of the Area". The NCAA initially consisted of various government officials and four Masaai elders. The Authority failed to function because there was little rapport between officials and elders. By 1960 a management plan was prepared for the NCA and revised from a sociological side by H.A. Fosbrooke, who was appointed conservator in 1962. A fine record of Ngorongoro appears up to this period (Fosbrooke, 1972).

In 1981, following the declaration of NCA as part of the World Heritage Site, a group of residents in Tanzania were commissioned to prepare a new management plan (Bralup, 1981). The focus of the plan was first and foremost the needs of the people, especially as there was an urgent need to dealing with the deteriorating food situation among the Maasai. Many studies were undertaken on pastoralists ecology, food security among the Maasai and on environmental degradation. The main points to emerge have been summarised as follows:

i.   Pastoralists, livestock and wildlife have coexisted in the area for over 2,000 years; pastoralist grazing and burning activities have helped to shape the area's present highly valued landscape.

ii.   Livestock numbers monitored for over 20 years have fluctuated but show no overall trend of increase.

iii.   Wildlife populations have undergone a dramatic increase over the same period, making the idea of adverse competitive impact of livestock dubious, if not untenable.

iv. Disease interactions between cattle and wildlife populations favour the latter.

v. There is no evidence to bear out suggested changes in vegetation composition, whether in pastoralist-occupied areas or in areas from which pastoralist stock have been excluded for 10 years or more.

vi. The NCA shows negligible erosion. Rates are lower than for all surrounding areas, despite the greater geomorphological and topographic predisposition of the area to erosion. (*Homewood and Rodgers* 1988).

The NCA highlights, among other issues, the contradiction between local versus international interests; the rights of communities versus the opinions of scientists; community participation versus bureaucratic decision-making; and planning and the consequences of the lack of it. The Ngorongoro case study amply demonstrates the dangers of blanket conservation prescriptions which bring great insecurity to the traditional dwellers of Ngorongoro. While there are no universal answers, there are options, but each one has a price. Obviously, there is no such thing as neutrality when it comes to conservation development.

Following the revised Management Plan, and the subsequent upheavals in Parliament over the treatment of the pastoralists and the activities of the conservation bureaucracy, a Parliamentary Commission of Enquiry brought changes and an uneasy truce for the Maasai to live in their ancestral homeland. The NCA demonstrates the necessity for caution. Human security and welfare are not necessarily ensured by orthodox conservation!

NATIONAL CONCERN FOR ENVIRONMENT

A striking feature about environmental stress is that in the Southern Africa region there is more and more awareness at national levels that something has to be done to redress past failures to take the environment into consideration. It is worth following the interesting evolution of various countries in tackling environmental and natural resources issues.

Zambia

In 1984, Zambia was the first country in Africa to have a national conservation strategy. In an address at the opening of the new headquarters of the United Nations Environmental Programme, President Kenneth Kaunda stated:

> "We in Africa are gradually coming round to appreciate that the protection of our continent's environment should be a priority.
>
> Development and environment are two sides of the same coin. Sustained development—whether it is based on industry or agriculture—is impossible unless it is also based on the sound management and efficient use of the resources of the environment....If unsound development is a major cause of environmental problems, then surely it follows that development itself must be analysed in order to determine ways and means of preventing its negative effects." (President Kenneth Kaunda, Nairobi 1984.)

Unfortunately, the Zambian NCS in its very short gestation period could not devote much time to analyse development itself.

In a summary introduction—"Why Zambia needs a National Conservation Strategy"—the following aspects are highlighted: deforestation, soil erosion, degradation of traditional pastures, pollution and poaching. the strategy recognised the need for environmental discussion to be included in development planning. Zambia is slowly breaking out of the sectoral approach. A national coordinating committee to implement the NCS has effectively been working for the last three years and is in the process of studying a decentralised approach to the NCS.

Zimbabwe

Zimbabwe was not far behind. Immediately after Independence, at the first National Conservation Conference of the Natural Resources Board in 1980, the Prime Minister promised the nation that, "my Government will implement and promote the World Conservation Strategy". Because of a proud record of environmental conservation and technical expertise, Zimbabwe decided to formulate its own NCS through an inter-ministerial steering committee under the auspices of the Ministry of Natural Resources and Tourism. By 1986 a document entitled *The National Conservation Strategy, Zimbabwe's Road to Survival* was completed. It was launched the following year (Republic of Zimbabwe, 1987).

However, major efforts will be required to convert a technical document into an action plan. Given the history of the country, in which over 80% of the rural population occupy the poorer half of the land, a technical approach poses too many problems. It has also been pointed out that the Zimbabwe Conservation Strategy cannot be implemented as a sole responsibility of the Ministry of Natural Resources and Tourism (Mascharenas, 1988). The Government faces a major dilemma. It cannot overnight demolish structures evolved during the colonial period and replace these with radically new forms of management without taking the risk of socio-economic instability.

Botswana

Botswana is the third country to embark on an NCS. The formulation has taken over three years, involving discussion of national technical committees, with both the public and private sectors. Above all, incorporating consultation has been incorporated at village level. The completed document will soon be presented to Parliament before it is enacted and ready for implementation. The question has been posed whether a relatively affluent country like Botswana needs an NCS.

Measured in per capita GDP, Botswana is one of the richest countries in SADCC. At independence in 1966, it was one of the poorest countries in the world, with a yearly per capita income of 50 pulas. Yet despite a population growth rate in excess of 3%, its per capita GDP in 1988 was 3000 pulas. Between 1973 and 1988, Botswana recorded one of the highest GNP growth rates in the world (Hermans, 1988).

Given this unique situation in Africa, is conservation relevant for a rich country like Botswana? The answer is an unrevocable "Yes", simply because the people of Botswana are dominated to an unusual extent by natural resource-based activities. It has been estimated that agriculture and mining would account for 44% of GDP in 1988/89 and around 80–90% of the value of exports.

Most of those employed in the informal sector are based in agriculture. The overwhelming dependency of Botswana on natural resources makes it critically necessary to develop a strategy for natural resources conservation—critical because mining, which contributes so much in respect of output, export and public revenue, is based on an exhaustible resource and provides very little employment. Agriculture is a far more dominant activity, especially in terms of employment. Unfortunately, under current practices, agriculture threatens to exhaust many of the renewable resources through overstocking and cultivation practices which cannot be sustained.

The long-term growth of Botswana requires the transformation of its present cattle-dominated agriculture. Although the ownership of cattle is extremely skewed, they are very important in the indigenous economy of Botswana. The livestock industry needs to address itself to the state of the range and the subsidized nature of its growth. To achieve sustainable development the economy will have to diversify from its present emphasis on livestock and minerals.

Security for Botswana's people arises from conservation because its natural resources are the only assets which are commonly distributed throughout the rural (and generally poorer) sectors of the population. A sustained improvement of living standards is impossible without conservation.

Ideas important to the NCS, such as sustainable development and diversification of its economy, are not new in the planning process in Botswana. However, the NCS emphasizes that in the development of Botswana, there is need to better understand the factors influencing the efficiency of renewable natural resource utilization. Also to be considered are intersectoral linkages and, given the prevalence of droughts, the intertemporal effects of these on economic activities.

There are difficulties in getting a systematic assessment of data and information on natural resources. For a number of resources there is little data—either because they are outside the formal economy as in the case of subsistence hunting, gathering and fishing, or because it has a low priority in government planning, as in the case of tourism. Recent estimates indicate that as many as 10–12,000 people still hunt for subsistence and that over 15,500 individuals found employment in fisheries, wood carving, wildlife farming, etc. A shift of emphasis on natural resources utilization offers some good possibilities for sustainable development.

Although Botswana is arid and semi-arid, the more dramatic potential possibilities for sustainable development lies in the fishing sector which hitherto has remained largely untapped. Apart from the 10,000 sq km of the inland delta of the Okavango and its flood plains, there are others such as the Chobe system, Linyati Swamps and Lake Ngami. The delta is exceptionally prolific as a breeding ground for a number of commercially important fish species. Nearly one-third of all small grants made through the Financial Assistance Policy (FAP) have gone to remote area fishing households. As a result, there are over 800 commercial fishing households, and yields have been improved by a factor of 20. In the northern part of the delta between 1982 and 1986, fish catches have increased from zero to 100,000 kilos!

Regarding wildlife, Government now seems to be becoming sensitive to the issue of the impact of livestock on wildlife. The creation of Wildlife Management Areas (WMA) will reduce pressure on the range by creating buffer zones. This and other measures will ensure the survival of both livestock and wildlife.

Therefore, it is important that one takes a longer and more imaginative view of the utilization of such renewable resources as grazing and arable lands, fishing and wildlife, forests, wetlands and water resources. Conservation viewed in this context raises hopes. It brings a message of development options, not a prognosis of doom.

Mozambique's Efforts at Coping with Environmental Stress and Fighting for Survival

The government and people of Mozambique face a daunting task in trying to meet the challenge of fundamental security-preservation of human life while facing the challenge of environmental stress. Given the paucity of information and even disinformation on Mozambique, this section portrays the difficult task ahead of that nation.

For several years Mozambican authorities have been appealing to the world concerning problems brought about by destabilisation. Little attention has been paid to this call for humanitarian assistance. Early in 1988, the Gersony Report commissioned by the US Department of State was made available to the press in Washington. Based on a three-month field visit to 42 sites in five countries and interviews with 200 randomly selected refugees, a chilling picture emerged. It is clear that the people and Government of Mozambique have horrendous problems to overcome.

In Mozambique the brutality and savagery directly or indirectly supported by the apartheid regime has been unprecedented. Nearly 300,000 people have already lost their lives, and infant mortality deaths in some areas have now reached appalling rates. Following Independence, improvements to human welfare had brought IMR down to 150 per thousand, but this now averages over 200 per thousand.

Innocent civilians in neighbouring countries have also been affected. Within about 15 months the number of Mozambican refugees in Malawi shot up from 70,000 to 500,000. In the Southern Nsanje District, the resident Malawian population of 150,000 had to accommodate 175,000 Mozambican refugees within a short time. The social and economic costs and the environmental stress on Malawi have yet to be assessed. Problems of inadequate agricultural land and poverty were already being experienced by Malawi even before the arrival of the refugees.

Environmental parameters

Despite its considerable natural resources, Mozambique continues to be one of the world's poorest countries. A prolonged war of liberation and continued fighting after independence to support Zimbabwe have meant that the country has simply not had a real opportunity to take stock of its resources.

Centuries of neglect have meant that little is known about Mozambique's biophysical resources. With few exceptions, the economic and political world has not been helpful to developing countries, and Mozambique has had to bear the brunt of this burden, compounded further by destabilization. As a result, 40% of Mozambique's GNP will have to go to debt repayment. The debt service ratio is 250, compared to an average of 179 for developing countries. Severe foreign exchange crisis and destructive activities by bandits have resulted in reduced services, reflected in the horrible demographic statistics.

Its colonial legacy gave Mozambique a very poor economic and social foundation. Its initial equity-oriented plans for social services brought remarkable progress. Health and primary education became increasingly common. For instance, an immunization programme was started in late 1976 from the Northern provinces and by early 1979 was 95% completed. The country has a fine record of political mobilization to bring about development. While the route taken was unconventional, the process has been educative and participatory rather than autocratic.

Mozambique shares with most developing countries the enviromental problems brought about by development, in particular, neglect of sustainability and management of natural resources. The majority of its people live in rural areas and depend on natural resources for their livelihood and well being. To get a proper perspective, it is worth taking stock of Mozambique's main environmental parameters.

The significance of the marine environment

Mozambique has the largest coastline in East Africa—2,700 kms. Together with the investments in marine infrastructure of Mozambique and other SADCC countries, the marine environment is crucial and must be the first priority for protection and management for three reasons:

Firstly, nearly three-fourths of the entire population of Mozambique is within 40 kms of the coast. Secondly, the marine areas are the habitats for a variety of marine life, some of which are Mozambique's major renewable resources. Thirdly, the coastal area and the shallow seas are also the sites of minerals and possibly hydrocarbons.

The dynamic relation between land and sea has to be known if the marine environment is to be sustainably managed. This necessitates collection of data, information and the examination of options and sound management.

The marine habitat

The continental shelf in east Africa is generally narrow, extending on an average between 15 and 25 kms. In Mozambique it is much wider, so that around the Bight of Sofala it is 145 kms wide. Therefore, within the 200 kms jurisdiction of territorial waters, Mozambique has nearly 120,000 sq. kms of shelf area.

An essential ecosystem is the mangrove. The protection of mangroves which are increasingly threatened is not a luxury. In the case of Mozambique they form one of the main habitats for shrimps and prawns. In 1986, prawns were the main national exports accounting for some 48% of all export earnings. Destruction of the mangroves could lead to drastic reductions of Mozambique's highly valued export.

Current estimates of the mangrove in Mozambique are around 84,000 ha. Knowledge of the composition of the mangrove forest is very much in its infancy. It is known that the *Rhizophora mucronata* is under pressure because it is cut for firewood and fuel for sugar estates. Pollution of water from oil refineries and ships, as well as chemical discharge, are likely to threaten mangroves. Studies of the relationship of shrimp productivity to the mangroves need to be conducted and then used in national decision-making.

There are over 60,000 artisanal fishermen involved in harvesting the inner seas of Mozambique. Fishing is important to balance the diets of the artisanal communities. With more information on technology, organization and infrastructure, the livelihood of these communities could be enhanced.

Offshore, Mozambique has not yet exploited the considerable stock of pelagic tuna which could easily double the present fish catch. Foreign ships are fishing in Mozambican waters but the levels of catch and whether they are sustainable are not known. Information about inland and freshwater fisheries is even more difficult to come by. Important as fishing and other marine products are to social and economic well-being, marine species which are rare, threatened or endangered must also be considered.

Environmental situation pertaining to agriculture

The degree of environmental change is to a large extent determined by the interplay of the type of agriculture, soils, terrain and the type of vegetative cover. It is estimated that in 1984 the family sector occupied some 4.8 million ha., out of which three million ha. were cultivated with annual and perennial crops. The state agricultural farms, mostly monocultural crops, covered less than 100,000 ha. As a generalization one could state that most of the major agricultural potentialities are to the north of the country. The two provinces of Zambezia and Nampula alone account for over 50% of peasant production.

Traditional systems of agriculture depend on fertilizers available from the virgin soil. Therefore, the human population density is a crucial factor. Extensive areas are needed, but as long as the population density is very low, the resulting slash and burn agriculture is viable and self replenishing.

The current destabilisation means that population densities have increased rapidly, and people have moved into new areas of differing carrying capacities. Environmental stress is caused by the reduction in the fallow period, and with increasing declines in crop productivity, even larger areas have to be opened or marginal land utilised, including steep slopes. The result is inevitable: soils extracted of their nutrients, the erosion of soil and the start of gullies, eventually culminating in desertification. In the drier areas, there is loss of the more nutritious grasses which are replaced by bush or less palatable grasses.

Livestock and the environment

The presence of tsetse in northern Mozambique has concentrated cattle and livestock in the South. The environmental problems are intensified for a different and additional reason. In the Angnia parts of Tete Province, the combination of a high concentration of people, the use of steep slopes and the presence of a high number of livestock has led to severe soil erosion. There is a multi-country approach which includes Mozambique, to clear flies using more environmentally protective methods (Stevenson, 1988).

Environment and the modern agricultural sector

The large-scale, monocultural estates are in the South. The cultivation of sugar and rice and the low rainfall have demanded that irrigation be practiced. This has meant major environmental transformation.

The cost to the environment of changes brought about by dam construction has been high. Dams have been considered mainly within the engineering framework, and some unnecessary drainage of swamps and marshes has taken place which is neither economically nor technically justified. Secondly, the changes in the hydrologic regimes have altered the conditions for the flora and fauna. Thirdly, a considerable part of the waters in some of the drainage basin originate in areas outside Mozambique and are contaminated with fertilizers and pesticides. The rich fisheries of Mozambique will suffer from this pollution. Fourthly, irrigation projects have spread malaria, schistosomiases, filariasis and diarrhoeal diseases.

The danger with monoculture of the modern sector is that it lays bare soil for part of the year so that it is exposed to the full force of weathering and destruction. In Manica and Niassa provinces, mixed farming is encouraged to stimulate conditions in which soil can recuperate (Muess, 1985).

Environmental issues related to petroleum and hydrocarbons

The Mozambique Channel has become a major highway for large-scale transfer of petroleum products from the Middle East to the developed countries in the North. Mozambique is therefore liable to get a larger share of pollution in its exposed marine territory. Ocean monitoring and surveillance is an extremely costly affair which Mozambique cannot afford.

There are ways of getting around these problems. Mozambique could sign the Regional Seas Programme and petition IMO for the creation of a non-dischargeable zone and simplify surveillance and enforcement of international standards.

More important, though, is oil discharge into coastal waters. The main refinery is in Maputo but so, too, are a large percentage of the motorized vessels plying in the country. The country is still in the early stages of oil pollution level monitoring and related activities. Activities and contigency planning for the eventuality of accidents must be brought to the forefront.

Currently, there is exploration for gas and other hydrocarbons. An EIA before any extraction projects are launched is essential. Fortunately, many of Mozambique's close supporters, particularly Norway, have considerable experience specifically related to hydrocarbons and the environment.

Environmental problems related to population and human settlements

There has been a tendency in recent years to blame the victims for their problems. Mozambique's present poverty, reflected by a GDP of US$ 152 per capita, is not a result of misuse of its resources. Rather, many of the environmental and resource problems are due to the destabilization. Given the size of the country and its population, densities are generally not excessive. Thus Nampula and Zambezia Provinces had a density of 29 and 26 per sq. km. respectively, during the 1980 census. However, the settlement pattern and the distribution of population are changing very rapidly.

Of major concern should be the pressure of population in the South. In one of the basic necessities, domestic fuel, there is already a severe problem. Fuel wood and charcoal have to be transported over long distances. Maputo itself has all the momentum to grow so that the problems will be compounded.

For instance, the population of Maputo city has grown astronomically. The population of Maputo has grown from about 755,000 to an estimated 1 million in 1988. This has put a massive burden on the natural resources because in desperation for domestic fuel, trees and plants have been cut down and thatch and building material are becoming very scarce. Because the majority came for reasons other than economic, there is an intense pressure on natural resources: land, water, domestic fuel. Since the social services were not geared for this influx, new hazards have been created. Waste disposal systems cannot cope with the huge additions, housing is makeshift and poses its own problems and is a fire hazard, etc. The quality of life has definitely declined. There is need to reverse these trends. Evidence from other capitals in Africa implies that Mozambique should redirect growth away from the capital city.

On a smaller scale, these problems are also found in the other provincial capitals. Since most of the important towns and cities are also ports, they experience stresses because two contrasting environments meet. Thus the sea is easily mistaken as a convenient dumping ground for waste and sand is all too easily scooped for construction purposes; and this inevitably begins the problem of environmental damage.

Natural hazards

Mozambique has had its share of problems from natural hazards. Severe hardships and loss of life occur not in the driest part of the country, e.g., the Limpopo Valley, but in the areas of higher and variable rainfall. South of Beira, the rainfall average is below 500 mm. South of the Save River about 90% of the terrain is less than 200 m a.s.l. and is covered by infertile sandy soils. Apart from low rainfall and variations, they also suffer from edaphically induced drought. To compound the problem, most of the rivers have their origin in some of the highest lands in Southern Africa so that they are subject to periodic flooding which cannot be controlled from Mozambique.

Mozambique should strive to be less vulnerable from climatic hazards. Therefore, as the country develops, it should obtain greater security by shifting emphasis to the North where the climatic hazards are much reduced (see Table II).

Given the above background, is an NCS relevant in a country like Mozambique which is desperately fighting for its survival?

The Government of Mozambique has put environment and conservation high up on the agenda for its rehabilitation and development. In order to obtain concrete results, the Prime Minister has appointed the Ministry of Mineral Resources to be the coordinating ministry, and the Ministry of Water and Construction will assist. The National Institute of Physical Planning will be the focal point for integrating activities.

Despite this situation, one is impressed with the determination of the people in Mozambique to bring back things to normal and to plan ahead. For a country that has not shied away from deciding on priorities that are important for its people, its more recent commitment to environment is not surprising. The liberation struggle was a success precisely because FRELIMO knew and operated in a rural environment. Future improvement of the standard of living and sustainable development will be dependent on the political commitment to the management of Mozambique's natural resources.

The war and economic difficulties have made the majority of people in Mozambique even more dependent than before on natural resources for survival. Given the pattern of its exports, natural resources are important even in the formal sector.

The fundamental problem of environmental stress and security in Mozambique

At present the fundamental problem of Mozambique is not environmental stress but destabilization, targeted on civilians and social and development structures. This leads to the following negative consequences on the environment and resources:

i) Population is increasingly being concentrated in a few strategic areas which makes heavy demands on natural resources, such as land, water, domestic fuel supplies, etc.

ii) Subsistence utilization of natural resources is being distorted and traditional sustainability is being replaced by a mining-for-survival syndrome.

iii) Systematic planning has become difficult.

iv) Major investments have virtually ceased to contribute to the national economy, best exemplified by the Cabora Bassa dam.

Given the political will by Mozambican authorities to do something about the environment and natural resource use, there is an excellent opportunity for intervention and support. This in no way underestimates the problems to be overcome. However, it would be callous to expect Mozambique to shoulder this immense responsibility alone.

TOWARDS SUSTAINABLE DEVELOPMENT AND SECURITY

Since the United Nations Conference on Human Settlement in Stockholm in 1972, which subsequently led to the foundation of UNEP, progress has been made in heightening the environmental awareness of governments. However, in sociopolitical terms, the needs, responsibilities and "rights" of human beings to a healthy and productive environment had to await the WCED report.

In the beginning of the 1980's, as a result of years of background work, IUCN (International Union for Conservation of Nature and Natural Resources), with the collaboration of other United Nations, bodies, came up with the World Conservation Strategy (WCS). On a global scale it drew our attention to the consequences of existing and frequently destructive ways of using our biophysical world. One of the major recommendations on the WCS was the need for action through

National Conservation Strategies (NCS). Three countries in the region have responded to these initiatives and others are soon to follow.

Conservation is fundamental to sustainable development. As defined in the World Conservation Strategy, conservation is "managing the use of environment and natural resources to ensure the maximum sustainable benefits for present and succeeding generations".

A National Conservation Strategy is an opportunity for a country to consider immediate and long-term basic material needs within its administrative and professional capacities. As a basis for discussion, programming and action will ensure that scarce technical and financial resources are not diverted. NCS's should be opportunities for creatively planning optimal present and long-term use of resources. In terms of security, the Brundtland Report has given an even stronger basis for considering the environment.

With "Our Common Future", the concept of sustainable development has taken a holistic dimension in the context of international and local political and economic responsibilities of all people. It is a seminal document in that it courageously brings to the forefront grey areas such as environment, development, women's rights and multilateralism.

For operational reasons and given that the majority of people in the Southern Africa region are dependant on natural resources, it is necessary to focus on the modifications to the biophysical environment and the use of resources to satisfy human needs and improve the quality of life.

Reduction of environmental stress has therefore become even more relevant and meaningful because the Brundtland report is not an agenda to *limit* growth, either now or in the future, but a dynamic approach which recognizes today's needs while trying to ensure continued use in the future.

It values human creativity which is more widespread, rather than restricting sustainability as a technical, sectoral issue. Against the background that resource depletion is avoidable through better management, this essentially means ensuring that life supporting systems continue to function effectively. The key elements for sustainable development are: conserving against loss of plants (trees, shrubs, grasses) and other living organisms and maintaining genetic diversity so as to keep greater options for the future.

CONCLUSION

In the Mozambique situation, in which national security is threatened, the survival of households is unpredictable. However, the state can undertake preparations to ensure that with peace, action can be undertaken to enhance the quality of life of its people, especially those in the rural area. From a resource potential, there will have to be a reorientation to the north and greater attention and priority given to the marine environment.

Even in areas of political stability, the Ngorongoro study reveals the complex nature of. resolving site-specific conflicts. Reduction of environmental stress for human security cannot be tackled as a sector. While the case of Mozambique shows that the disadvantaged have to bear an even greater burden, one must not lose track of the marginalised people in all the SADCC countries.

To deal with meaningfully, problems of environmental stress must be considered in socio-political and development terms. Given the political support and the elaborate process of consultations at grassroot level attempted by Botswana, there is hope that the amelioration of environmental stress will be a participatory activity. By shifting the approach from things to people, from a sector to a more integrated approach, the environment/natural resource issues begin to point to development opportunities closer to the realities of rural life, introducing greater self-reliance, opportunities for employment, etc.

Given a period of peace and stability, the Southern African countries can collectively come closer to resolving many of the problems of environmental stress. This will ensure national security and allow communities to become custodians and partners in sustainable resource use. The grey areas brought to the forefront in the Brundtland Report will have to become even more prominent in the Southern African countries. There is an unpredented opportunity to make this happen in the Southern Africa region.

REFERENCES

Andersson, D. and Grove, R. (Ed) 1988: *Conservation in Africa, People, Policies and Practice.* Cambridge, Cambridge University Press, 355pp.

Arhem, K. 1981: Maasai Pastoralism in the Ngorongoro Conservation Area: Sociological and Ecological Issues *Bureau of Resource Assessment and Land Planning,.* Dar es Salaam.

Arntzen, F.W. and Veenendaal, E.M. 1986:*A Profile of Environment and Development in Botswana*. Institute of Environmental Studies, Amsterdam, National Institute of Development Research and Documentation, University of Botswana, Gaboro..e, 172 pp + appendixes.

Bhalotra, Y.P.R. 1985b: *The Drought of 1981–1985 in Botswana*. Department of Meteorological Services, Ministry of Works and Communications.

Botswana, Republic of/UNDP 1988: The National Conference on Strategies for Private Sector *Development Francistown*. Ministry of Finance and Development Planning/UNDP, Gabarone, 178pp.

Bralup, 1981: A Revised Development and Management Plan for the Ngorongoro Conservation Area, *Bureau of Resource Assessment and Land Use Planning*, University of Dar es Salaam.

Bruce, J.W. 1985: *Land Tenure Issues in Project Design and Strategies for Agricultural Development in Sub-Saharan Africa*. Land Tenure Center, University of Wisconsin-Madison, 195pp.

Brundtland, G.R.O. 1988: *Our Common Future*. The World Commission on Environment and Development, Oxford University Press, Oxford, 383pp.

Christiansson, C. 1981: *Soil Erosion and Sedimentation in Semi-Arid Tanzania..* Scandinavian Institute of African Studies, Uppsala, 208pp.

Fosbrooke, H. 1972: Ngorongoro: The Eighth Wonder, Andre Deutsch, London.

Fosbrooke, H. 1938: Ngorongoro Crater and Serengeti National Park, Paper Topic 5, Management of Protected Areas for Sustaining Society, Forthcoming paper in International Congress on Nature Management and Sustainable Development, 6–9 December 1988, University of Groningen, Groningen.

Gersony, R. 1988: *Report on assessment of Mozambique Refuqees*. Maputo, Mimeo c40pp.

Hjort, G. 1982: A Critique of ecological models of Land Use, *Nomadic People* No. 10.

Homewood, K. and Rodgers, W.A. 1988: Pastoralism, Conservation and Overgrazing, In: Anderson and Grove, *Conservation in Africa. People. Policies and Practice*.

Leach, G. and Mearns, R. 1988: Bioenergy issues and Options for Africa, IIED, London, 214pp.

Mascharenas, A. 1983: Ngorongoro: A Challenge to Conservation and Development, *Ambio*. Vol.XII Nos 3–4, 1983.

Mmusi, P.S. Hon. 1985: Closing Statement: The Chairman, Vice President and Minister of Finance and Development Planning, SADCC—1985, *Proceedings of the 1985 Annual Conference Held in Mbambane*. SADCC Gaborone, 180pp.

Mturi, A.A. 1981: *The Archaeological and Paleontological Resources of the Ngorongoro Conservation Area*. Ministry of National Culture and Youth, Dar es Salaam.

Sen, A. 1981: *Poverty and Famines*. Oxford University Press. Oxford, 257pp.

Wisner, Ben. 1988: *Power and Need in Africa: Basic Human Needs and Development Policies*. Earthscan Publications Ltd, London, 351pp.

World Commission on Environment and Development:*Our Common Future*. Report on the World Commission on Environment and Development, Oxford University Press, Oxford, 383pp.

Zambia, Government of the Republic of, IUCN. 1985:*The National Conservation Strategy for Zambia*. IUCN Gland, 96pp.

Zimbabwe, Government Of The Republic of, 1987: *The National Conservation Strategy: Zimbabwe's Road to survival*. Ministry of Natural Resources and Tourism, Zimbabwe, 36pp.

TABLES

Table I. *Ngorongoro Conservation Area. Human/Livestock Ratios 1954–1987*

Year	Human Population	Cattle	Ratio	Stock Units	Ratio
1954 (Grant)	9,321	22,263	13.1	208,263	17.6
1974 (Frame)	13,178	123,613	9.4	155,114	11.8
1977 (Frame)	16,705	110,584	6.6	159,310	9.5
1987 (IUCN)	22,637	137,398	5.1	206,087	9.1

Source: Fosbrooke.

Table II. *Recent natural disasters in Mozambique*

		Provinces affected
1976	Cyclone Dainae	Inhambane, Gaza, maputo
1978	Zambezi River floods	Sofala, Manica, Tete, Zambezia
1982	Cyclone Justine	Nampula, Zambezia
1984	Cyclone Demoina	Sofala, Inhambane, Gaza, Manica
1984	Drought	Manica, Gaza, Inhambane, Sofala Tete, Nampula. Over 3.5 million affected
1985	Drought	Manica, Sofala, Tete, Nampula. Over 1 million affected.
1986-88	Partial drought	Several regions throughout Mozambique
1988	Floods	
	- Limpopo river	Gaza
	- Pungoe-Buzi River system	Sofala
	- Zambezi River	Zambezia
1988	Tropical depression Filao	Zambezia

www.ingramcontent.com/pod-product-compliance
Lightning Source LLC
Chambersburg PA
CBHW080608270326
41928CB00016B/2963